CW01656218

Dedicated to my family

Preface

Over the last nine or ten years, so many people have asked how we came to live and work at such a beautiful farm and why an ordinary, healthy family chose to build and run a Farm Centre for people with disabilities.

There's no simple answer, so I usually mumble something about it being a long story, and then change the subject.

But whenever those questions are asked I find myself looking back on our strange journey, from a little caravan in a scrubby field to the natural splendour of Blaen Wern Farm, and smiling at some memory of the past.

I remember the characters we've met along the way, wonderfully eccentric people who have worked the land and shared our love of the countryside, people who for one reason or another have brought joy and humour into our lives.

I remember too the antics of our animal friends, the horses and dogs who played such an important role, each one adding love, loyalty, laughter and, whenever possible, chaos to our daily chores. And I remember the cattle and sheep, enthusiastically joining in the disruption and thwarting the best of plans.

And I'm reminded again of the natural world that we are privileged to know and call our home. The majestic hills and lush valleys, wooded slopes and sparkling streams. And western sunsets, vivid and ever changing shades of gold and red against the darkening blue of an evening sky.

So perhaps I can indulge my fundamental love of Nature and tell the story through them all, people and animals set in the glorious rural countryside of West Wales.

This then, is my way of answering those questions. A simple abridgement of how the destruction of our home placed us on the road to Blaen Wern, how the farm prompted the building of the Centre and how we subsequently found ourselves involved in the lives and needs of disabled and sick children.

Chapter One

Arrival in Wales / The New Land / Caravans / Snowy the Rabbit

Country people are noted for their cordial hospitality. Visitors are welcomed as friends, and their needs are met with a generosity forged from generations of hardship and understanding. Imposition is a different matter altogether, and this conglomeration had all the ingredients of a full scale invasion.

A bright yellow lorry led the way, its thudding diesel engine objecting noisily as the driver carefully negotiated the sharp bends and steep hills of a sparsely populated region in West Wales. Its open back was piled high with a bizarre assortment of bits and pieces all securely held in place by a spider's web of nylon rope. In its passenger seat a small fair haired boy held a white rabbit sitting happily on his lap, and towed behind was an ageing fourteen feet touring caravan. Following closely in its wake was a woman in a small car, with another boy and his young sister staring excitedly out of the windows.

An approaching Land Rover pulled sharply onto the grass verge, its mud spattered wheels spewing chewed up grass onto the tarmac surface before the selection of four wheel drive took it clear of the advancing apparition. Its driver rolled his eyes skywards, appealing to some unseen rural deity to deliver him and his quiet backwater from the disruptive forces that undoubtedly must accompany such a motley horde of foreign intruders. He acknowledged the proffered thanks with a polite but hesitant wave, and a weather beaten farmer leaning on his farm gate bit hard into the stem of an empty pipe as he stared in disbelief, his gaze following the gaudy object of his attention until it disappeared from view around the next bend.

The convoy pulled off the narrow country lane, squeezed past a broken gate and bumped slowly along a rough farm track that ended abruptly at the unfenced edge of a scrubby looking field. The vehicles stopped and the children tumbled out, running without pause up the adjacent hill in a frenzied attempt to burn off surplus energy conserved to bursting point during a long, slow journey. The adults emerged more slowly, walking

stiffly to and fro to stretch aching limbs and embracing the unkempt landscape with tired eyes. They exchanged knowing smiles that betrayed inner feelings of satisfaction and fulfilment. The woman disappeared into the caravan to put the kettle on.

We'd arrived. It was 1977 and the first day of our new life.

Our destination had been duly noted. Rumours would soon begin to circulate as rumours do, and the confusing contrast between the well groomed appearance of Irene and the children and the free rein given to my complete indifference to either fashion or convention would no doubt add further fuel to the speculation. Anxious eyes would be watching closely for the first signs that would give credence to their not unnatural suspicions - cannabis, poppies or any other offbeat plants decorating the landscape and thrusting their pseudo happiness up from the Welsh soil as a first step to the ultimate goal of establishing an 'alternative lifestyle' or 'new world' that were all the rage at the time with college dropouts who thought they had the answer to every problem on the planet and were determined to prove it - provided of course that their giros arrived on time.

It was just a few hours earlier that we'd left the security of a little English village that had been our family home for untold generations, a village desecrated by the building of more and more houses for more and more commuters to create more and more wealth. Slowly but surely fields and common land had been eaten away in the cause of progress, indigenous wildlife dwindled and local countrymen became strangers in their own land.

Apart from being the only home I'd ever known, there was nothing particularly special about it, a few farms, deciduous woods and a thousand acres of common land. It was rather disparagingly referred to as Hungry Hill by those who weren't a part of its close knit community, a reference to the flints found in abundance at the top end of the village. It was those same flints that had attracted the first residents many thousands of years ago. Its only claims to any sort of limited fame were the presence in considerable numbers of stone age implements and even older fossils, and the dubious reputation of probably having more pubs per capita than is likely to be found in any other similar location - which, together with an apparent ability to produce very large families, probably speaks volumes about the after work activities of my ancestors.

In spite of its unpretentious simplicity, or perhaps because of it, I found the escalating erosion of my childhood playgrounds difficult to accept, and for years I'd felt claustrophobic, unsettled, dissatisfied and, although I don't remember anyone saying so, was probably becoming difficult to live with. We'd talked many times of moving permanently to West Wales, but we'd shied away from the big and emotional step of selling our home. I'd built it in my spare time during the year before we married and its sentimental value had grown with the three additions to our family. The final decision was delayed time and again - perhaps because we lacked the courage. I suppose we could have been accused, with some justification, of hovering between a dream and a plan.

Yet most weekends we travelled the two hundred odd miles to the land we'd grown to love for its natural and unspoilt beauty and for its friendly and homely people. I felt very strongly that our future was there. It was where we were meant to be. Sometimes we stayed for weeks on end, looking for a new home that we weren't even sure existed beyond the very clear mental picture of a farm in the hills that had good productive land, a lot of mature trees and pure spring water.

We needed a push to make us take that final and irrevocable step. A gentle nudge would have probably been enough but, exercising the customary undemocratic authority that we've all come to know and expect, fate didn't ask for my apparently irrelevant opinion. What we got was a vicious metaphorical thump in the back that stunned us for some weeks and forced a number of drastic and fundamental changes to our lives.

A gas explosion split our home apart destroying or seriously damaging everything in it. By some miracle no one was at home and only the building suffered. Even so it was an experience we could have happily spent the rest of our lives without. Finding ourselves homeless with three young children to care for was not a situation we would have cared to recommend to others - unless they were particularly crazed and dedicated to high profile experiments in the dark and mysterious world of masochism.

It was in the days before index linked insurance was the norm, and a couple of weeks had passed before the financial implications hit us. Only then did we realise that the house was grossly under insured and that ten years' work, together with the money to buy our farm, had gone, literally in a flash.

There was an inevitable period of shock and disbelief of course, but the most vivid memories are of a sudden change in attitude and sense of values. It all sounds a bit over emotional and possibly even sanctimonious now, but it was real enough at the time. The futility of accumulating material possessions for personal gain forcibly struck me and I wondered why I hadn't understood that before. We were all learning the hard way that they're only temporary and have no real value. The most precious parts of my life were safe and well, and I realised that nothing else really mattered. It was a traumatic way to learn a simple lesson, but we all came out of the experience closer and stronger than ever before. It was a period that was to influence all our lives in a variety of very positive and constructive ways.

There was no choice but to start all over again, to build another home that would also be a saleable asset and a springboard to the future. But this time it would be in Wales. If this setback had any sort of practical outcome, it was to fix our eyes even more firmly on that farm in the hills - wherever it was.

So we sold everything of value that was still in one piece, and wouldn't be needed for the task ahead, to raise what we hoped would be enough cash to start us off. There were no misplaced illusions, we knew that there would be precious little money to spare and that spending would have to be restricted to a very tight budget until we were established and earning again.

As a first step we bought thirty four neglected acres that hadn't been farmed for years, with no house, no water and no electricity. There was no barn and no fences, and only a few stone walls remained to suggest that it had once been a well managed part of a bigger farm. It was rough to say the least, and the nearest road wide enough to accommodate two vehicles side by side, if they were very careful, was about a mile away. Geographically, it sat at the head of a valley leading down to Cardigan Bay, on the ancient route between Aberaeron Harbour and Strata Florida Abbey high up in the mountains, and its only tenuous link with a wider world was a network of narrow country lanes between the dairy and stock farms and the 'major' road which led to a central source of supplies and the market towns.

Most of the fields were ploughable, although it would never be the most productive of land, and at eight hundred feet above sea level the growing seasons were relatively short. But it was snug enough, and the vicious autumn winds were kept at bay by gentle, protective hills on either side, except at one point near to the road where a clear view of the sea exposed the land for periodic battering by the north westerlies.

We had to turn those barren acres into a viable agricultural holding before we could move on. It was a challenge. That was how the holding got its name, from a chance remark one day that "mighty oaks from little acorns grow", we called it Mesen Fach - translated into English that becomes Little Acorn.

Looking back now, it's easy to understand why local people kept themselves at a safe distance for a week or so. We did get an early visit from the children of a neighbouring farmer who called round to sell us their eggs and milk. We weren't quite sure whether they were sent to effect a cautious introduction or to dig out a bit more information about our background and intentions so that it could be converted into juicy scandal at the local chapel, so we said nothing that could be distorted by repetition. It briefly crossed our minds, somewhat uncharitably, that they were possibly there just to make the most of a commercial opportunity, but as the minutes ticked away and they showed no indication of heeding their parents warnings about being too friendly with the doubtful and as yet unproven newcomers, we decided that their visit had more to do with innocent country curiosity than with anything more sinister. Whatever their reasons, they broke the ice locally by spreading the word that we appeared to be harmless enough, even if the general opinion still remained that we were at best a couple of sandwiches short of a picnic.

They could have been right of course, and looking at the situation again twenty years on, with the benefit of a bit more wisdom and a lot more caution, I think they probably were. But at the time the thought that we were taking on too much didn't occur to us. We'd thought long and hard about the task ahead and were reasonably well prepared in a practical sense as well. We'd received outline planning permission for a new farmhouse before completing the purchase, and if I carefully demolished what was left of our old home, we'd have a sizeable and very useful assortment of materials that would make building it much less of a burden on the finances - and we'd resisted the temptation to add to our modest finances by selling a couple of diggers that had provided part of our income in the past by being hired out to contractors, in the belief that they would more than pay their way in the months and years to come. To transport it all to Wales, I'd bought the yellow lorry at a GPO surplus vehicle auction, the main attraction being the hydraulic lifting gear so that I could load and unload on my own. We planned to give a whole new meaning to the term 'moving house'.

We spent the first few weeks trying to acclimatise ourselves to the new conditions and environment, and organising food supplies, schooling and the thousand and one other things that needed organising before there would be any hope of starting to bring plans to fruition. Our caravan was a long way short of being ideal. We were cramped and sometimes cold and damp, although with memories of all the pressures we'd left behind still fresh in our minds, the peace and tranquillity was exquisite. The children were understandably apprehensive about joining new schools but they didn't let that stop them from enjoying the open spaces and making the most of the freedom to explore them, wisely treating the whole episode as an adventure to be savoured and not an experience to be endured.

We discovered that there are disadvantages to living in a caravan for more than a week at a time that we'd never thought of before. We'd always been used to casually strolling from one room to another in a house built for a growing family, and living like sardines in an undersized tin took some getting used to. Cooking was something of a nightmare, trying to satisfy the demands of the hungry horde with limited space and miniature implements, and a constant worry in case the children should somehow fall against the little cooker or grab a hot pan when passing. Whenever possible they were banished to the great outdoors until the dangers had passed and the results of Irene's labours were ready to be devoured with the usual enthusiasm.

Bedtime was even worse. There was never enough room for beds to be left intact during the day, so they all had to be made up just before the children were called from their play to wash and sleep. With the exception of a small area in front of the sink and cooker, where we could stand upright, any movement from one end of the caravan to the other at meal and bed times was by a sort of stooping crawl with the occasional trip and scramble thrown in for good measure. It came as something of a relief when the preliminary squirming was at an end and everyone was either eating or bedded.

Privacy became almost non existent and the smallest juvenile blemish was exaggerated by the close attention of a brother or sister hoping to derive some entertainment from a newly erupting spot on an otherwise faultless face as it glowed and throbbed away to the shameful mortification of its owner. Us adults had long since passed the age of either subjective or objective interest in isolated facial, or any other, flaws but we still didn't escape the cruel attention of our loving children who delighted in highlighting other imperfections that we'd never before noticed.

Snowy the children's rabbit wasn't the least bit upset by any temporary deprivations because he didn't have any. We'd brought his hutch with us, an up market affair that I'd put together in a weak moment, with a spacious sitting room complete with picture window and stairs leading to a very comfortable bedroom in which he'd built his own cosy little nest. We quite envied him. But in spite of his well stocked larder and palatial accommodation he wasn't satisfied and couldn't wait to have a closer look at the new scenery. Within a few days he was escaping almost at will, repeatedly gnawing his way through the plywood backing to go for an afternoon stroll and happily sample the tantalising greenery in our rabbit version of the Garden of Eden.

The back of his hutch quickly became transformed into a patchwork of rapidly executed and increasingly bodged repairs that bore all the hallmarks of anything but a master craftsman, and when we reached the point of having to put patches on patches defeat appeared to be inevitable

and we left the door open so that he could come and go as he pleased. It pleased him to reject the grandeur of his purpose built home in favour of a nest at the end of a self dug mini warren under the caravan, and the only time his freedom was subsequently restricted was when he wandered too far from base and was in imminent danger of becoming a predator's supper.

I was impatient to emulate Snowy's domestic bliss and get started on the real work of providing a more suitable family home, although it was patently obvious to all of us that it would have to wait until our initial accommodation had been expanded and improved. The trouble was, we weren't yet certain how temporary our temporary home would have to be. It was just remotely possible that we'd have the house built in a little over a year but theory and practice seldom turn out to be the same and realistically it was more likely to be two or three. Even the briefest of cursory glances at our little caravan was enough to confirm that it would be totally unsuitable for a period that long, especially since it was already mid September and we had the full ravages of winter to look forward to.

There was a natural concern that spending too much time and money at a very early stage could have an adverse effect on the already tight budget and create problems later on. Yet whatever we chose to do would effect all of us, one way or another, for many years to come, so it was right that we should all agree. As I recall, the resulting discussion developed a growing tendency to be eloquently enthusiastic on the subject of investing in one of the quality mobile homes of massive proportions that always seem to be available at the end of the tourist season, and grudgingly monosyllabic on the alternative. It was more than a little worrying for the head of the house and keeper of the family purse.

Financial logic is about as popular a topic of conversation with the young as homework and bed time, and sometimes the only way to be certain of a solution agreeable to those of more mature years and claimed superior judgement is by being a bit devious. With that in mind I had little option but to camouflage totalitarianism with a much more attractive veneer, and exercise an assumed parental right to interpret a succession of particularly submissive grunts and mumbles from the children as wholehearted support for the proposal that we should overcome the problem as cheaply as possible and put up with temporary inconveniences for the benefit of the long term aims.

I'm not sure how, but I somehow managed to convince them that the advantages would be considerable, without actually making any firm promises that I wouldn't be able to honour. Thus the democratic process triumphed again and the motion was eventually carried unanimously.

So during the next few weeks I made numerous more journeys in the lorry, first to acquire and collect two more caravans and then to transport the diggers and other tools and timber that would be needed for the first stages of the planned building schedule.

Our home became transformed into a comparatively spacious living 'unit', five times its original size, with three dilapidated caravans, each one parked at right angles to the last and forming three sides of a square, and looking something like a wagon train anticipating immediate attack. The two smaller ones became our bedrooms, one for the children and one for us. A big twenty six footer that I'd bought in Hampshire for forty pounds scrap value was used as a sitting room and a kitchen. Towing it to the west coast turned out to be a never to be forgotten journey that took a full twenty four hours on the road, and almost everything that could go wrong did, including a wheel falling off at about the half way mark.

In spite of the problems, the effort was worthwhile and whilst the resulting additions may have been a bit rough and basic, they were a welcome improvement for everyone. The extra space made a big difference to our day to day lives and we were ready to face the winter with a lot more confidence.

Chapter Two

Finding Water / Laundry / Television / Sindy the Kitten / Bath Time / Generator

There was a negative side to multiple caravan living. In a very short space of time the simple process of walking from one caravan to another proved to be something of an ordeal as the patch of grass between them responded to the first few showers of rain and a never ending procession of scuffing feet by slowly turning into a sea of mud. Even by our low standards, putting on wellington boots to go to bed was something less than acceptable.

The only feasible solution that wouldn't cost a fortune was to adjust the caravans to bring their floor levels to the same height and to build a suspended floor in between. The end result was in all honesty a bit of an abortion, but the pioneer in me saw such an improvement that I covered the whole thing with a boarded roof and lined it with alkathene sheeting. Viewed from the road it was a long way from being pretty, and it leaked a bit during heavy downpours, but at least the floor was clean and free of mud - and, more to the point, so were we.

We had the same problem on the ten yards or so of grass between the caravans and the track. The obvious solution was to lay a gravel pathway, but since we didn't yet have any aggregate on site and something had to be done quickly I used some of the surplus timber to make up some rough duck boards. It worked quite well, in a very basic sort of way, but every time a gumboot clad body dragged itself from the deepening mud on each side and squelched towards the caravans, half a bucketful of sticky, black Welsh soil fell from its temporary host and waited for the next pedestrian to obligingly spread its greasy consistency even further over the boards. So walking its length during wet spells turned out to be a bit of a lottery, and it was advisable to tread with great care and to take very small steps.

Water had to be collected in five gallon containers from a local farm, not only for drinking and cooking but for washing as well. In common with most people, we'd always had a never ending supply permanently available on demand so it came as a bit of a shock to discover just how much we needed. In spite of treating its unnecessary use as something approaching a major crime, we always seemed to be either close to the point of running out or with barely enough to make a cup of tea. Most of it went on washing and the lack of personal hygiene facilities was perhaps the biggest drawback to our primitive way of life. If there were ever any quiet moans they would almost certainly be about that, and as if to rub salt into the wound the big caravan had a bathroom, complete with a conventional white bath that without water was less than useless and served only as a constant reminder that however often we washed we never felt clean. Our own supply of water had to be the next, and urgent, priority.

We did have a stream flowing right through the middle of our land which at a push was suitable for washing but it was a long way from the caravans, and in any case we'd have had to be pretty thirsty to drink it. It rose in the hills not far away so it was fresh enough, but the rust coloured tinge of eroded sandstone was a bit off putting. Herons paid us a visit now and again so I suppose there must have been fish in it somewhere but we never saw any, although the children discovered that it wasn't completely barren when they unwittingly disturbed a sizeable, and extremely irate, eel. We decided to give it a miss and look elsewhere.

The topography of the general area suggested that there should be a spring somewhere on the land and a location map from the deeds distinctly showed one not too far away. But some old surveys that go back over a hundred years are not always too accurate, and an initial brief search showed nothing that was immediately obvious. Then, a few hundred yards from the caravans, we came across a shallow ditch, well camouflaged with weeds and running along the edge of a field. Following its length up the slope of the land we found that it started abruptly and for no obvious reason, and using our bare hands we unearthed a small square of rocks covered by a slim flat stone that hadn't seen the light of day for many decades. Underneath was a little pool, with crystal clear water gently bubbling to the surface. It was a promising start, and it could well be useful at a later date, but for our immediate needs it was at too low a level and would need to be pumped over the hill to the caravans.

There seemed to be only one way to find water closer to home. Over the years I'd used dowsing rods to find underground cables and pipes as casually as the more technological apparatus is used today. I'd even amused myself by locating electromagnetic changes in the earth's crust, and pondered on their cause and effect - but I'd never looked for water. Now seemed a good time to have a go, and coincidentally provide any passers by with the final proof that the newcomers had not only lost their marbles, but more than likely didn't have any to start with. So gritting my teeth and trying hard not to look in the direction of the road I spent the next hour or so wandering backwards and forwards across the fields with an intense look of concentration on my face, elbows tucked firmly into my side, arms protruding forward like a wooden soldier and holding two thin copper rods in outstretched hands - which of course were totally invisible beyond a range of about fifty yards and therefore to any passing onlooker.

I never discovered if this obscene ritual was witnessed by our neighbours, but by the time I'd covered every square foot of about ten acres of land I'd lost complete interest in trying to protect my already tarnished image. In any case I was by then too flushed with success to care what they thought. Practising the theory that 'you don't know what you can do unless you try' had paid dividends, and my hastily fashioned copper rods had obediently indicated more than one possible site, each one now marked with a wooden stake ready for further investigation. I was very pleased with myself.

However, since my thirsty and expectant family was patiently watching and waiting during the whole operation I have to admit that the successes were blighted by one or two dismal failures, each one reinforcing their already well established scepticism and providing further proof that my dowsing skills were not yet honed to perfection. The mental art of differentiating between metals and water had proved to be more difficult than I'd at first thought. It was a deficiency brought home to me a number of times as I enthusiastically dug up a multitude of ancient and broken plough shares, which may well have been of interest to local historians but didn't contribute greatly to our current needs.

I finally exposed one or two reasonable sources, and one of them was only about five yards from the caravans. Its output was modest to say the least, but it was good enough for our immediate needs and after just a few minutes work with one of the diggers we had a pure and convenient supply of domestic water.

As a bonus Irene had a brand new example of one of the earliest types of washing machine - clean water from a hole in the ground and a hard rock. The elbow grease she had to supply herself, although with an astonishingly accurate insight into the domestic needs of a wife and mother I'd anticipated that other tools of the job of keeping the family reasonably clean and presentable may be required, and had put one or two labour saving devices on the lorry. The wringer had been a welcome modern development in its day and was infinitely preferable to the old fashioned mangle that decorated every country outhouse in my childhood. It was light and required a minimum of effort to turn the handle, and with the addition of a purpose made bath to catch discarded water underneath, which I'd also remembered to include, considerably simplified one of the more arduous processes in the whole operation. Irene of course was delighted with my loving thoughtfulness - or at least she didn't say she wasn't.

The tumble drier came from the same period, in the form of a line slung between two trees on the exposed side of the hill. It was a spot chosen with great care in order to ensure that the clothes were well and truly dry before the return journey to the caravans had been completed, thereby contributing to the degree of physical fitness that was essential to our adopted way of life. Most of the tumbling unfortunately came from Irene as she scampered across the fields in hot pursuit of airborne laundry baskets caught up in sudden gusts of wind. Now and again she wasn't fast enough and they disappeared forever, although one escapee did eventually come to light four months later, trying to take root in the middle of the adjacent forest.

The problem of ironing was solved in the same considerate way, and half a dozen flat irons that hadn't seen the light of day since world war two were rescued from the dark and dismal recesses of a family loft to earn their keep yet again. It took Irene some time to develop the necessary skill required to bring the cast iron to exactly the right temperature on the gas stove, by which time every one of my shirts without exception was artistically decorated with at least one brown triangular scorch mark somewhere on its beautifully laundered surface. Strangely, the mistakes only seemed to appear on my clothes, a coincidence I've never quite been able to understand.

With laundry difficulties removed, a regular supply of food and drink assured and our general living conditions elevated almost to a point just below squalor, there was time to move on to other, more progressive, matters, and a lack of storage space was another problem that was in need of urgent consideration. We were still the proud owners of a battered collection of temporary sheds that hadn't been sold off before our move, simply because the chances of finding any sort of serious buyer, apart from vendors of kindling wood, were nil. Faced with either buying a box of matches or transporting them to Wales, we'd pinned our hopes on them surviving the journey and then hanging together long enough to protect our surplus belongings until the house was built.

So while Irene played with her new toys I made yet more journeys in the lorry to rescue them from further deterioration and desecration by the elements and, since they were bulky but not all that heavy, topped up each load with a ton or two of salvaged furniture, more building materials and anything else that I thought might come in handy and that would conveniently and safely fit into an empty space without falling off en route.

Putting them up could be likened to piecing together a giant jig saw without the benefit of a guiding picture, except for one or two chalk marks long since rendered indecipherable by the Hampshire rain together with a hazy recollection of their previous form dredged up from the depths of a suspect memory. I finally convinced myself that all the bits were in the right places, although standing back to view the end result did nothing to boost my confidence, so I stopped looking.

There were probably a few more raised eyebrows from passers by, but they served a useful purpose and soon became full to bursting point with fire damaged, but still usable, furniture together with so many unfamiliar odds and ends that we began to wonder if someone else was quietly filling our sheds with their own surplus chattels under cover of darkness. Only the thought that no one in their right mind would want to store furniture in those conditions convinced us that it must all belong to us.

Mesen Fach mansion was beginning to take on some sort of rough shape, and by the time a few of the more familiar ornaments and pictures had been strategically placed in the caravans, the ramshackle accommodation had become welcoming and homely. Pride of place was taken by a tiny black and white television that we connected to a spare car battery. It was an important psychological addition although it wasn't without its problems. Car batteries are notoriously unreliable and there were a few howls of disappointment each time the power ran out at the crucial point of one of the children's favourite programmes, the picture getting smaller and smaller until it disappeared completely in the middle of the screen.

It's presence apparently also had the effect of filling a social void in the children's lives at school, allowing them to participate in the deep and meaningful discussions on current subjects of international importance that followed cultural and academic transmissions - which I found a little confusing since every time I saw them in front of the little box it was displaying cartoons or children's soaps.

It was about this time that Sindy the kitten joined our family. It wasn't our idea, but a close neighbour was desperate to find her a home and as newcomers we were the obvious targets. It appeared that Sylvia's husband had brought the cat home as a present, either because he'd forgotten about her feline phobia or possibly because he was trying to tell her something in a particularly peculiar and devious way. Whatever his reasons, we ended up with a cat to look after at a time when it was as much as we could do to look after ourselves. In spite of our initial reluctance we all quickly fell in love with her and she earned her keep by adding to the growing homeliness of our scruffy abode as only cats can.

Sindy was very affectionate, and the children made sure that her frequent requests for a fuss were seldom met with refusal. Being spoilt with a surfeit of attention, she naturally developed an unshakeable conviction that the sole purpose of clothes was a convenient means of reaching the security and comfort of shoulders, and that laps were specially created by our Maker for cats to sit on.

One of our early visitors was a lady canvassing for votes in the forthcoming local elections. She was quite obviously a town person but in order to fool the peasants had dressed for the occasion in a smart tweed suit and brogue shoes, believing, with a characteristic lack of understanding and awareness, that country people spend their lives similarly attired. She would have been better received had she appeared in gum boots rescued from the muck heap and smelling like the inside of a cow shed. Like many politicians she also suffered from a combination of patronising self esteem and verbal diarrhoea, which in this case were unfortunately aggravated by an extraordinary ability to spout absolute rubbish with all the misplaced confidence of an idiot. Our insistence that we had no moral right to vote at all because we hadn't lived in the vicinity long enough to fully understand the issues in hand was totally ignored, and we were left with the impression that it would be perfectly acceptable provided we voted for her. None of this particularly enamoured me to either the politician or to her politics.

Sindy, though, ignored my prejudices and took an immediate liking to the lady's lap, arriving at her chosen destination with one well timed leap that left her balanced athletically on her hind legs and thrusting a moist, purring mouth well into the depths of the political right ear. The resulting scream and rapid departure left us in no doubt that our budding councillor was not well suited to representing farmers, and more than likely had the same affinity to all animals as she obviously had to cats. Sindy's credibility immediately rose to new and dizzy heights in our household, and for a long time afterwards she could do no wrong.

With the help of our new feline addition, we were as domestically settled as we were going to be for a while, but time was rapidly passing and there were still a lot more jobs to be done before the building work could be started. For a start all available materials had to be collected from our old home, and that meant quite a few long journeys which would take me away for at least three days at a time. Those weeks turned out to be probably the most difficult periods for the rest of the family.

Mark and Paul were fourteen and thirteen, so were old enough to accept some of the responsibility and to look after their mother and sister Nicola, a role they both accepted with manly dignity and adult confidence. Even so, there was little they could do to help on one black day that they all still vividly remember.

It was Sunday, a day of rest when rural communities attend to the stock, go to Chapel and return home to slumber. It is also a time of false security, when all those inanimate objects that we tend to take for granted try the patience of a saint by choosing the same day to give up the ghost.

Cooking and heating were the first to go. The bottles of butane gas that decorated the outside of the caravans reached the end of their allotted span and collectively expired with a last desperate gasp and flicker of flame. The fire got the blame, for over reacting to a drop in temperature and gobbling up the last remaining energy reserves with even more enthusiasm than usual. Next came the television and a couple of twelve volt lights. The car battery decided to join in the fun and go on strike until a power boost had been arranged via a mains socket and twelve volt charger. The last straw was saved until after the decision had been made to walk to the phone box and contact me on an emergency number that had been arranged for such eventualities. As the caravan door was opened, a freak wind came from nowhere and ripped it from is hinges.

It all ended happily at the end of the day, Sylvia's husband arrived like a knight in shining armour, and within two or three hours the door was replaced, the battery charged and a shopkeeper called from his weekly siesta to provide more gas. It was the country way. A stranger came to help and left as a friend.

Although we soon learnt to anticipate those sorts of setbacks and to take all necessary precautions, living in caravans and being without electricity and water on tap did have its obvious drawbacks. But it's surprising how quickly we all adapted to a more basic lifestyle and how soon we realised that such luxuries aren't really all that important.

Mind you, there were times when we allowed our minds to flit back to an earlier, more self indulgent, existence - and not only on laundry days. Preparing to have a bath wasn't quite as simple as we may have initially anticipated, and it invariably prompted an added determination to get the house built as soon as was humanly possible. It was a long and laborious job that made the actual bathing operation a quick and necessary function rather than an enjoyable relaxation.

Water had to be carried from the spring in buckets and heated in an old butane gas boiler that we'd had enough foresight to bring with us. Then more water was carried to get the bath to the right temperature and give enough depth to make any sort of all over washing a reasonable possibility, usually about four or five inches. It was crucial that we literally took the plunge without delay. The consequences of dawdling were learnt by bitter experience. Steam rising from the hot water in the boiler immediately condensed on the low ceiling to form a growing number of freezing cold droplets which hung there sadistically, preparing to fall onto the bare back of whoever had decided that a bath couldn't be put off any longer.

So after a quick scrub, and an even quicker rinse down, there was a lot of hopping about from one foot to the other as an overworked towel slowly massaged circulation back into reluctant action and restored some sort of dull feeling to an otherwise numb body.

Later on, we really moved up market when I fitted a manual centrifugal pump to lift water from the spring direct to the kitchen, or anywhere else within reach of the plastic hose, which in my infinite wisdom I had decided was an indispensable part of the overall plan. I'm tempted to boast that it was a very technical and modern piece of equipment, but in fact it was primitive to say the least. It was operated via an eighteen inch wooden handle that had to be worked in alternate directions through about forty five degrees, and obviously the faster it moved the faster the desired result was achieved. The general consensus of opinion was that it was probably harder work than carrying the buckets. On the plus side, if you were quick, and lucky, the body heat generated by fifteen minutes of strenuous pumping was retained until after the bathing operation had been satisfactorily completed.

Electricity was always going to present us with a bit more of a problem. The mains supply was some distance away and it was, we were told, a major undertaking to extend the high voltage line to our land. So the earliest we could hope to have a supply laid on was in six months time - and we took that estimate with a pinch or two of salt.

Meanwhile we were still ferrying car batteries to and from a local farm for charging, in a sometimes vain attempt to keep a few dim lights working and to head off the tendency of the television picture to disappear into its own interior. It was an uphill battle and something had to be done.

Right on cue, word filtered through on the grapevine that a hospital in Wiltshire was selling off its standby generator in favour of a newer model. I reasoned with characteristic acumen and confidence that if it had been used to shed light on the subject of a surgeon's scalpel and to cast even the faintest of glows over countless adjacent wards, it was reasonably safe to assume that it would light the way to our caravan bedrooms without too much fumbling and stumbling across the timber staging.

Money was the problem, or our lack of it. The laws of supply and demand dictate that equipment that doesn't often come onto the market can sometimes go for a lot of money, and it was anyone's guess what sort of silly sums could change hands if circumstances had pushed a prospective buyer to a point of desperation.

So it was more in hope than expectation that I posted off a supremely optimistic and ridiculously low offer calculated to suitably match our finances, not really expecting to hear any more about it. To say that we were surprised when our offer was accepted is a gross understatement. We were heart stoppingly amazed, and in next to no time half a ton of energy production was on the back of our lorry and on the way to Wales before they could change their minds.

There was great excitement the following evening as we prepared to switch on the Mesen Fach illuminations. I'd spent the day setting it all up a hundred yards away from the caravans, with a solid wall of straw bales on all sides to muffle the rhythmic thump of an oversized diesel engine, and laying power cables to the consumer unit in the kitchen. And to spoil ourselves even more, I'd wired up the start and stop buttons so we didn't have to go outside to operate it.

All lights were switched to the on position and we were ready for the big moment. The engine gave a few token coughs and splutters before, responding to the combined will power of five anxious onlookers, it burst into life and disgorged its ten kilowatts of energy, lighting up the whole place like an overgrown Christmas tree.

Now we had everything we needed. However rough and basic it may have been, we had warm dry shelter with fresh water and electricity laid on, and television for the children. Irene had become good friends with Sylvia who was a regular visitor to our modest abode, and the children had made their own friends locally and through school. We were fairly well established and a great deal more relaxed.

Those feelings would only be temporary of course, and without the visual encouragement and incentive of a growing house, there's little doubt that dissatisfaction with our accommodation would have crept in at some point in the not too distant future. But, surprising as it may seem, I still look back on that comparatively short period as the happiest and most contented in my life.

CHAPTER THREE

Building the Barn / Cocky the Bantam / Dai and Megan / Haymaking / Glyn

Getting the necessary council approval for the detailed plans of the house took a lot longer than I'd hoped so to smarten the place up a bit, and also to give ourselves a more tangible sense of permanence, I fenced off a couple of the fields and then started work on the stock building, even though we didn't yet have any stock to put in it. Well, not unless you count Cocky.

Cocky was a bantam cockerel. You know the breed - beautifully coloured, firm of flesh and admirable athletic machines. Not this one though. It has to be said that he was the most dreadful looking apology for a bird that is ever likely to walk this earth. He had absolutely no value and even less use, and he was the first 'livestock' to grace our new holding, given to the children by a neighbour who was doubtless glad to see the back of him.

Watching his pathetically ill conceived form arrogantly strutting around the caravans, incessantly talking to himself and taking out his predatory spite on innocent worms unwise enough to pop up for a breath of fresh air, provided irrefutable evidence that his unfortunate physical deficiencies were easily surpassed by a mentality well below the basement level of the average stupid chicken. The worth of his previous owner's generosity became tinged with an element of doubt and there was a sneaking feeling that he probably didn't stop chuckling for a full week. Ironically though, it was Cocky's shortcomings that justified his continuing presence, providing the perfect qualities for his adopted role of farm clown.

But, give credit where it's due, he was a born survivor. His first brush with death came early in his life with us, on the night of a severe north westerly storm that blew up from Cardigan Bay.

As usual he'd roosted on Irene's wringer stored in one of the temporary sheds, and as usual we'd shut the door behind him. By the early hours of the morning the storm was upon us, the wind was howling and the caravans began to rattle and shake ominously. Sleep was out of the question and we could only lie there until the worst was over and a fitful doze became a reasonable possibility.

At about six o'clock we were brought back to reality by an almighty gust of wind, followed by the sound of snapping timber and a thundering crash. A quick glance out of the window told the whole sorry story. Cocky's shed had disappeared, ripped from its base and scattered across the fields in countless directions, smashed to pieces by the unstoppable force of a freak wind. But there, where the shed had been a few minutes earlier, with a confused look on his face but still sitting on his wringer, was Cocky.

Later on, we bought some hens to keep him company, and of course to give us fresh eggs every day. But it was on the strict understanding that broody hens would not be allowed to sit. The possibility of lots of little half bred Cockies contaminating an otherwise noble breed was unthinkable and out of the question.

Being given the new status of leader of a family of females immediately went to his head, and the size of his already oversized ego multiplied overnight. The following day saw him strutting about like the Lord of the Manor, clucking a natural leader's instructions and calling his harem to feast on some minuscule titbit that he'd managed to laboriously scratch from the soil. The hens showed remarkably good taste and totally ignored him.

Regrettably, for Cocky, that was the pattern for the future. However much he clucked and crowed he couldn't attract one single follower and spent much of his time racing to catch up after his wives had taken off in a different direction leaving him still strutting purposefully forward and talking to himself.

Even marauding foxes ignored him. They depleted our chicken numbers from time to time, but they never touched Cocky. In a way that was the ultimate humiliation. But we understood their reasons. Such a scrawny little specimen would hardly have provided enough for a mid morning snack and I doubt that they even gave him a second glance.

The intelligence and endless persecution of foxes has passed into folklore, but their unspoken criticism of our livestock did nothing to enamour us to their cause and was an unnecessary reminder that we had a long way to go. A few fleeting glances at one or two quality cows and sheep in the area had already told us that, and I don't suppose we were the only ones to notice.

Neighbouring farmers were still hovering in the background, watching our progress to see what sort of a mess we'd make of things and waiting for signs of any unwanted foreign culture being thrust into their jealously guarded way of life. We'd been in that position ourselves so we understood their reservations and were prepared to wait, although in my experience, newcomers into a rural village are often not considered to be worthy of their place in the community for at least twenty years, and even then they're only accepted with some reluctance by native born locals.

But as the stock building neared completion people began to take us more seriously and we got to know them a lot better. It was an unexpectedly pleasant surprise to discover that we were being accepted so soon, and I made the most of it by admitting an ignorance in almost every local farming sphere and picking their brains unmercifully about Welsh methods. I probably came close to overdoing it and pushing my luck a little bit too far, but they always seemed more than pleased to give me the benefit of their lifelong experience and I was more than pleased to receive it. Their help was invaluable and I owe them a lot.

The parents of our first young acquaintances were particularly helpful. They farmed something approaching two hundred acres on the other side of the valley, and their land extended almost to the lane bordering our own, so we were close neighbours even though they lived about a mile away.

Dai was a quiet unassuming man, carrying on a long established family tradition on a farm passed down to him by his father and his father before him. Megan his wife was born and raised in the next village and was the driving force in the family, regularly expressing her thoughts and opinions to Dai in a torrent of Welsh, intermittently sprayed with a varied assortment of carefully selected and emphasised English words that could be easily recognised and understood by everyone - except perhaps very young and well brought up children. On very rare occasions Dai felt the need to retaliate, and then it was best to wade through the blue cloud and quietly leave by the nearest exit.

When they weren't discussing their differences they were charming, friendly and very hospitable, plying their guests with a mountain of home baked bread and cakes liberally coated with delicious home made preserves selected from a seemingly limitless larder, all of which would have to be washed down with a mug of weak, milky tea that I never did quite get used to.

We soon found ourselves fitting in with their work schedules and providing extra pairs of hands when they were most needed, helping out with shearing and dipping, and at haymaking time by either stacking bales in the fields or building bale ricks in voluminous sheds.

They tended to be social events, and we were either shedding buckets of sweat endeavouring to play our part in the well organised shambles that was apparently absolutely essential to the satisfactory completion of the annual tasks, or in the house being encouraged to stuff ourselves silly on roast meals of massive proportions until the thought of a return to our posts necessitated an adrenaline powered surge of will power that with a bit of luck would offset an all consuming lethargy brought on by the mandatory over indulgence.

Our colleagues for the day took it all in their stride, bending and stretching with no apparent ill effects and already talking about the next stop so that they could do the same thing all over again. It appeared that one of the necessary criteria for voluntary farm work was having a stomach like a dustbin. We did our best to keep up but found it impossible to shrug off a distinct feeling of inadequacy.

Our own early haymaking was less demanding of our digestive systems, and dehydration was only a remote possibility instead of a racing certainty. We only harvested about nine or ten acres and the lack of either muck or bagged fertiliser on the land was reflected in a yield that fell a long way short of the overflowing barns of our neighbours.

Unlike them, we didn't gorge ourselves every couple of hours but we did have more time to take a lot of short breaks, usually to sit on a bale and take yet another swig of Irene's excellent elderflower champagne, made with flowers picked from one or two elder trees growing at the far end of our land. It was deliciously refreshing, and potent enough to fill the summer air with the shrieking sounds of childish laughter and send hay covered bodies toppling backwards over stacked bales to roll giggling down the hill in a tangled web of arms and legs towards the new fence. The children couldn't get enough of it, but in view of such unseemly, playground behaviour by the adults, their intake had to be severely restricted.

We bought our machinery as a job lot from a farmer in a nearby village whose farm was somewhat lacking in good management and going downhill fast. He was trying to rectify a substantial financial deficit that had long since passed the stage when recovery was even remotely feasible, by buying and selling all sorts of farming implements, the majority of which had degenerated into various stages of decrepit uselessness.

Glyn had discovered somehow that we had a lorry with hydraulic lifting gear and a winch, and since his second greatest problem was collecting discarded implements from obscure corners of remote farms the news was delivered straight from Heaven. I was suddenly the best thing since sliced

bread. My big mistake was refusing any payment, and to this day I remain convinced that he bought up every bit of abandoned machinery in the area just to make full use of the free collection while it was available.

But he was a likeable man who tried hard to be honest with everyone, although faced with such an escalating liability at the bank he sometimes found the temptation to stretch the truth to its limit far too much to resist. It all depended how he felt on the day, and whether an unopened letter from the manager was sitting on his mantelpiece.

Unfortunately Glyn was his own worst enemy. He wasn't really cut out for selling, and however hard he tried he still remained as transparent as an open door. On the rare occasions that a potential customer showed even a passing interest in some item of his machinery, the mere thought of making a profit raised his blood pressure to dangerous levels which in turn prompted a sympathetic upwards surge in the pitch of his voice. The higher the potential profit, the higher his voice. Those who knew him simply negotiated downwards until his voice was back to normal, at which point they'd reached the best price possible, and promptly clinched the deal.

We selected our implements with a great deal of care. It still wasn't the best available but the price was right and it kept us going for a year or two, even though hay making was frequently interrupted by an urgent need to get the spanners out again. The baler was the main offender and needed constant adjustments to make the bales the right size and density, each stop being marked by words learned from Megan and with Glyn's name being taken in vain. But we managed somehow and never failed to bring in a decent crop. It did let us down once when a collecting arm broke and rain was forecast, but we didn't have to ask for help twice. Within minutes Dai had arrived and quickly finished the job with his own tractor and baler - and then stayed on to help us get all the bales under cover.

We seemed to have been accepted into the community, and it was comforting to be reminded for the second time that in an emergency help would always be on hand. Yet, as far as is possible, we were still masters of our own destiny and had the undisturbed peace and independence that was important to us.

Chapter Four

Trout Fishing / School Centenary / David's Bike / John / Thelma's Mother

It was a good time for Nicky, one of those rare periods in life when everything you do works out better than expected. Unfortunately, for every winner there's usually a loser somewhere, and Paul chose the wrong time to introduce his little sister to the delicate and highly skilled world of trout fishing.

Up in the hills, just a few miles away and well out of sight of marauding tourists, was a beautiful lake fed by the purest of mountain streams and set in a volcanic depression untouched by man since the last ice age. It was full of healthy, naturally reared trout. Few people bothered to fish there, in those days the existence of such places was taken for granted by residents in the more remote parts of the country, and the greed that now decimates Nature's resources was unthinkable.

It was a full day's excursion and, armed with sandwiches and drinks, they left early so that Paul had time to put a couple of trout into his bag for supper before the serious work began of showing his sister how it was done. That, of course, is what older brothers are for, to pass on acquired knowledge and experience for the benefit and understanding of those of lesser years that they care for.

And so we waited for their return with some expectancy, frying pan prepared for action, and ready to offer consolation to a young lady who had foolishly stepped into a man's world only to discover that she was well out of her depth.

They returned in silence, trudging down the driveway with Paul in the lead and despair written all over his face. It appeared that conditions weren't right for fruitful fishing. The heavy rain we'd had a week ago last Thursday had washed silt down from the hills above, upsetting the feeding conditions and disturbing clumps of weed that had deliberately propelled themselves onto his hook, masking the bait and frightening away the monster that was about to offer itself in unconditional surrender

- and anyway the water was too warm. Parental words of consolation flowed.

Nicky listened to her big brother's mournful narrative in respectful silence and with sympathetic understanding, her hands buried deep in the sandwich bag hanging loosely from around her neck. They knowingly looked at one another in the silence that followed and a broad grin spread across her face. Slowly and deliberately, and with a gesture no doubt mentally rehearsed many times on the journey home, the content of her bag was proudly disclosed to reveal the most beautiful trout you will ever see, its considerable length drooping forlornly over both sides of her outstretched hands.

Paul's face told the whole story. There was the expected disappointment and the look of someone who has suddenly dropped two or three rungs on the ladder of life, but there was also a look of unselfish pride. He was at the wrong age to say so, but he was quietly pleased for his little sister who, because of her willingness to try, thoroughly deserved her brief moment of glory - even if it was at his expense.

Nicky had also settled in quite happily at the little village school. It was a typical stone and slate construction built during the last century to provide the offspring of local farmers with some sort of rudimentary education that would enable them to carry on the rural traditions of their ancestors with an added acumen and awareness. Ironically, it was that increased awareness that eventually proved to be counter productive by prompting the mass exodus of young people from the farms to the centres of industry and the big money. The school was now run by a bright young headmaster who somehow managed to maintain a successful blend of the fundamental rural culture with progressive methods of teaching aimed at the preparation of his young charges for higher schools and colleges, and for the wider world beyond our small village.

1979 was a special year in the school's history, a year that celebrated a century of service to the community and provided a good enough reason for the staff to involve the children in organising commemorative displays and concerts to mark the special event. The festivities ended with a service at the local chapel, and the children contributed to the beautiful unrehearsed harmony of the congregation with well rehearsed renditions of classical Welsh compositions played on guitars and recorders. We found the whole thing quite an emotional experience. This was followed by speeches from one or two people of advanced years who could well have been present at the original opening ceremony, the content and meaning of which passed lyrically and unchallenged straight over our heads. Nicky's command of the Welsh language had improved daily, but I'm ashamed to say that our ability to understand a word that was said was somewhere close to zero. Those present who occupied positions of authority and respect appeared to understand our dilemma and did their best to help. When the time came to pass round the collection box we were told what was expected of us in perfect English.

The boys took a little longer to adjust in spite of having each other for support. They were at the wrong age to change schools, which was something we'd worried about from day one. It helped that Dai and Megan's children were at the same school and they both became regular and welcome visitors to our home, Rhian because it had recently been brought to her attention that she was a young lady undergoing pubertal hormonal changes and she therefore found herself attracted to the two boys like a moth to a flame, and David because he could at last associate with boys of his own age on his home territory, and incidentally show off his new 125cc Trial Bike, purchased by his parents as a birthday present but which doubled up on the farm as a faster and more comfortable means of rounding up the sheep.

Mark and Paul were quite envious of the new bike, although David generously shared his good fortune and allowed them to take turns at riding it at top speed around an obstacle course that they'd all marked out across the fields. It was essential to the maintenance of due respect for the owner that David should be the most proficient rider and clock up the fastest times, an elite position he enjoyed without fear of serious opposition for some days. Practice, however, is a great leveller, and as the boys' confidence grew it became necessary for David to take more and more unnecessary chances in order to preserve his status. He had yet to learn that the best way of making a fool of yourself is to try too hard to impress, and a stray rock that wasn't there on the previous lap took it upon itself to act as tutor.

There followed a sudden and unplanned parting of bike and rider, and David found himself flying through the air with all the grace of a duck on ice before overtaking the machine on his hurried way to a solo landing on the grass.

A quick check on the rider revealed no broken bones, but to be on the safe side we carried him inside to look for signs of internal damage and to assess the degree of shock. The boys displayed great faith in the value of our medical judgement, and left mundane life or death matters to us while they disappeared across the field in great haste to make sure that the bike wasn't damaged. We were finally satisfied that David had suffered no real harm apart from a badly dented ego and drove him home, leaving the boys still 'discussing' who should have the privilege of riding the bike across the fields back to Dai and Megan's farm.

The incident ended happily over mugs of Megan's weak, milky tea, and David suffered no physical after effects, although a stern lecture from his parents and a few days of derisory comments from his sister were inevitable consequences that were just waiting to explode in our wake and further depress an already cowed young man.

The piercing scream of a high revving bike became less frequent after that, and the only other sounds that found their way into our secluded little haven came from sheep and cattle and a few tractors at harvest time. The one exception was a stream of loud expletives which periodically echoed across the valley during the late evening or after nightfall, and which rose to a furious and sustained crescendo before tailing off to an ominous and intriguing silence. Then we knew that John's tractor was playing up again.

John lived and farmed on his own, and his past was always a bit of a mystery. He was a man better known for his physical strength than for his knowledge and wisdom, and for reasons best known to himself preferred to sleep during the day and work at night. He was sometimes to be found on a neighbouring farm during daylight hours, helping to bring in the hay or wrestling with some intransigent beast, but in five years I only met him three times.

His daytime absence presented an enormous psychological opportunity to his neighbours, an asset they unfailingly milked to the last drop. Since he could seldom be found to defend the charge that his bull of doubtful pedigree had made friends with a neighbour's aristocratic cow, or some other heinous rural crime, it followed that he was always available to take the blame. Whatever it was that had gone wrong, it was always John's fault.

We never discovered how the tractor reacted to his venomous assaults, although its continual refusal to work for more than an hour at a time may be a clue, but if the same treatment was adopted to encourage his plants to grow, they must surely have withered and died - which no doubt would have prompted even more nightly expletives.

It was already apparent that our new village home was much the same as villages everywhere, with a predominance of unconventional and original characters as diverse in their makeup and habits as it's possible to imagine, and who collectively give the countryside that unique and attractive social flavour. It was perhaps predictable that newcomers with little regard for convention should home in on one or two of the more outlandish cases, people who could no doubt be considered by urban dwellers to be extreme and a bit odd. But they were genuine and honest people and it was our privilege to quickly come to know them as friends.

One of John's few friends was a typical example. She was a local lady who also farmed on her own and was as unorthodox as he was. Thelma was brought up surrounded by affluence but had left home as a young lady to live a simple, and often basic, life on the farm. In spite of a frequent need of parental help to bail her out, she had never lost her determination to be independent, although by the time we came to know her it was looking increasingly like a lost cause.

Thelma's mother was the only stabilising influence in her life, although at ninety odd years of age there was no longer very much that even she could do to resolve the multitude of personal and financial problems that plagued her daughter's day to day existence. She was a charming lady whose pampered background had instilled an unexpected humility and a warm caring nature, so we were saddened to learn one day that she had finally succumbed to the ravages of more than nine decades and had peacefully expired.

It fell to Thelma, as the only surviving offspring, to attend to the necessary arrangements and to make sure that the old lady had a funeral befitting her status as a loving mother and benefactor, and reflecting her social standing as the widow of a senior government diplomat. It was a tall order. These up market affairs can cost a lot of money, and as usual money was the one commodity conspicuous by its absence. Her dilemma deepened when quotations started to arrive from the black suited gentlemen who make such an excellent living from the inevitability of dying. Thelma's budget could just about cover the charges levied by the crematorium, but the cost of transporting her mother's mortal remains by conventional black hearse was exorbitant and she refused to pay it.

Country people may appear to lack a certain finesse when compared to slick city theorists, but years of fighting the lethal combination of bank managers, bureaucrats and the weather has ensured that they are, if nothing else, resourceful. That was why a respected old lady of impeccable breeding, and an enthusiastic supporter of her local church, came to suffer the ultimate indignity of being loaded into the back of Thelma's battered estate car, to rest serenely on clumps of discarded hay and straw left over from its last foray into the seething cauldron of local calf sales, and securely tied down with best quality baler twine. Her companions for the journey, carefully parked next to her, were two plastic containers holding the water that would be needed to top up the radiator at regular intervals en route to compensate for substantial losses through a split hose, and an hydraulic jack which had been included in case the balding tyres ruined the whole well planned operation by picking up a mischievous pin.

There was an argument put forward that anyone in need of such transport is past caring about mundane matters like the make and condition of vehicle used, and one car would do much the same job as another. I'm not totally convinced that her mother would have wholeheartedly agreed, but without moving into abstract and emotional realms there wasn't really very much left to say.

It would have worked without a hitch, and very few people would have ever known about it, if only Thelma had remembered to fill up with petrol before setting off. She only noticed the gauge needle stuck firmly at the lowest extremity of the red warning mark as she drove sedately away in her mother's company for the last time.

It wasn't a good start but it could have been a lot worse, and a short diversion to the nearest garage should have been the only hiccup in an otherwise faultless day. Another couple of teaspoons of petrol in the tank and it would have been. As it was, there was still nearly half a mile to go when the engine gave a few token coughs and splutters before giving up and stopping completely.

The situation was now becoming desperate, and even if Thelma had invested in some sort of rescue insurance, which was a ludicrous and futile thought, the attending mechanic would be likely to cast more than a casual glance at her silent cargo and beat a hasty retreat to the security of the nearest police station. There was no option but to start pushing, and hope that the limited energy available would be enough to convey car and passenger as far as the petrol pumps, and that she wouldn't be stopped along the way by some awkward busybody wanting to ask a lot of silly and embarrassing questions.

Luck was with her for once, and half an hour later a madly perspiring Thelma stumbled weakly into her car to glide gently over the last few yards down the slope and onto the garage forecourt. The pump attendant showed remarkable restraint and self control, calmly delivering the required fuel and trying hard to give the impression that customers carrying deceased persons in the back of an estate car was something he saw every day of the week. But then he knew Thelma, so perhaps he really wasn't too surprised. Bumping into him a few weeks later, he casually asked if it was Thelma's mother in the back of the car, because he couldn't think of anyone else it could be.

The rest of the journey passed without further event, and the dear lady was duly delivered to her penultimate destination where family and friends had gathered to say their fond farewells. I suspect that the subject of the day was more than a little relieved that a line could now be firmly drawn under the whole episode, and that she could at last gather together the tattered remains of her once glorious dignity and rest in some well earned peace.

BP
X43

Chapter Five

Livestock Marts / The Calf Man / Lammy the Ewe / Quackers the Duck / Annabel the Pig / Pippa the Collie

With the stock building completely finished, the time was about right to buy a few head of livestock of our own before starting work on the house. It was one of the secondary objects of the exercise and would be a way of getting some money back on the grass and hay as well as tidying the fields up a bit.

There's one thing you can bet your life savings on in a stock rearing area, whatever day of the week it is there will be a livestock mart somewhere within relatively easy travelling distance. We were spoilt for choice, but completely lacking in experience. One was much the same as another to us. The wise move would have been to ask someone's advice, but we'd come this far without showing the slightest hint of bowing to the demands of wisdom, and there seemed no reason why we should start now.

What we hadn't reckoned on was being spotted as newcomers before we'd got anywhere near the sale ring, and therefore viewed with barely concealed suspicion by everyone present. The vendors hoped we were buyers and nodded and smiled every time we looked in their direction, willing us to join in the bidding. The buyers wished we hadn't come at all, sitting as close as possible to the unfortunate animal being paraded for inspection, with eyes twitching and fingers in spasm and occasionally looking in our direction in case we had any ideas about interfering and costing them money. We didn't look at the auctioneer in case he misconstrued our intentions and we ended up making friends and enemies all on the same day. Other participants in the day's goings on couldn't have cared less one way or the other but were dying to know why we were there anyway.

We kept them guessing for a few weeks, until we thought that the reason for our periodic appearances had lost its initial mystique and appeal, and we could seriously consider entering the fray without either encouragement or animosity. The ploy didn't work. Acknowledgement of our interest from the auctioneer was the signal for the owner of the next beast in the ring to give us a hundred and one reasons why we should buy his stock and no other, and for the regular buyers to raise the stakes to an unacceptable level with a flurry of bids calculated to force us out of the running. The lucky vendors beamed at us in gratitude.

It took a lot more time and perseverance before we were allowed to make a purchase, but even then it wasn't clear exactly why, and we spent the next day or two vainly looking for a health problem that everyone else had spotted except us.

Every now and again the frivolous moods of the buyers suggested that they had either dined well in the local hostelry at the expense of the auctioneers or had been plied with generous quantities of wines and spirits by optimistic and devious vendors in the hope that such generosity would be remembered as their stock was put through the sale ring. On those exceptional and infrequent red letter days they appeared to acknowledge that we were relatively harmless and extended the hand of tolerance by not bidding us out of sight and permitting the purchase of another auction lot.

In that hit or miss fashion we began to build our livestock numbers, first by buying the children their own calves to rear on the bottle, and then by adding a few more to make the feeding routine worthwhile.

Each new addition was given a name that was supposed to match its peculiarities and characteristics, from Freda the Friesian to Houdini the Hereford, who was always escaping from his pen, and Bully, who you'll be surprised to learn was a bull calf, through a few other thoroughly pampered beasts with stupid names, to Funny Girl who wasn't the most handsome cow you'll ever see and so became a family favourite.

Our experiences at the marts were to stand us in good stead in the years to come but having elevated ourselves onto the first rung of the livestock ladder there was no longer the time, or the inclination, to be continually subjected to the patronising condescension of the financially mighty, so we subsequently employed a buyer to find what we wanted at the right price. He was a man of about sixty who had spent a lifetime buying animals and in his later years did the job more for love than for money. The children called him "the calf man" and he was a regular visitor, sometimes to deliver a new purchase but quite often it was just a social call.

There was only one thing "the calf man" liked better than a glass of whisky, and that was two glasses, a character trait discovered during one memorable visit when he felt an urgent need to sample a previously unopened bottle that we'd brought with us. He had delivered a yearling heifer and was about to leave when the heavens opened and bombarded all and sundry with some of the biggest hail stones seen this side of a golf course. The nearest place of refuge was the caravans and with a reaction that belied his age he headed straight for them at top speed. We had a clear vision into the future, but there was no time to warn him. The momentum generated by a sprint start carried him unerringly over six feet of intervening mud to land heavily and with all his considerable weight on the edge of the first duck board, which was by then liberaly covered in slippery balls of ice, cunningly concealed by the slimy deposits of two day's welly boots. His feet accelerated even faster, and the failure of his body to catch up left them with nowhere to go except upwards and over in a perfect execution of an almost complete somersault. The inevitable result was an ego bashing splash down, face first into the mud.

The remainder of his journey to the caravans was a little slower and much less eventful, and having scrubbed him down as best we could the production of the whisky bottle seemed the only right thing to do, although whether it was to calm his nerves or to ease our consciences I'm not really sure.

It did occur to us that our purchases of cattle could be at an end and that we should perhaps consider moving into sheep, but "the calf man" eventually left for home none the worse for the experience, and possibly even a little happier. The incident was never mentioned again, and we often wondered if the 'medicine' had in some way wiped out all memory of the reason for the need.

It could have been a coincidence that, at about the same time, Dai and Megan gave Paul a pedigree Suffolk ewe lamb that had been abandoned by its mother. He called her Lammy - not as corny as it sounds, it was his nickname at school and a quick glance at our surname will tell you why.

Like all sheep, Lammy hated being left on her own and followed us everywhere. She had a very good appetite to the point of being downright greedy, and would push and shove for a bottle of milk whenever she felt the slightest bit peckish. She was spoilt rotten, and generally made a nuisance of herself until she got her own way. If we were indoors and dared to forget the time, a series of head banging thumps would rattle the caravan door until someone acknowledged her rights and produced the overdue sustenance.

At night she slept with the calves, butting them out of the way until she'd found the softest bed in the building and then looking disdainfully around the pen with a smug expression of contentment and satisfaction. We began to understand why her mother didn't want her, and as it turned out would have saved ourselves a lot of problems over many years if we'd shut her in a pen of her own right from the beginning.

We didn't know much about sheep apart from our experiences with Lammy, they were almost non existent where we'd come from, but we were anxious to learn and we knew from the past that our neighbours would be good and willing teachers. So we bought twenty of the draft Welsh Mountain ewes that were born and reared on the mountains to the east of us and sold off at the end of each summer to farmers on lower land.

They're excellent, hardy sheep but, because they've been used to the freedom of vast areas of mountain, they're also notorious fence breakers. We had the perfect answer to that little problem. Lammy knew the boundaries and she'd keep them in check, just as she did the calves.

Like so many foolproof plans, it didn't work. No amount of coaxing and pushing would persuade her to mix with those strange woolly creatures, and she just stood wistfully by the field gate staring in the direction of her beloved calves, bleating pitifully for their attention. There was only one sensible thing to do, but we didn't do it. Not for the first time, or the last, we were swayed by the pitiful eyes of an unhappy animal, and the gate was opened.

We were fairly sure where to find her, and for just about the only time in her life she was predictable. There she was, grazing happily in the middle of a dozen or so young cattle, occasionally glancing in our direction to make sure she'd made her point.

With the introduction of sheep and cattle our new home began to take on the shape, appearance and smell of a Welsh agricultural holding. Things were coming together quite nicely and I was almost ready to get started on the house, but there were one or two minor jobs that needed to be put out of the way first. Without Lammy to keep them under control, the sheep were constantly pushing their way through to the better land so more new fencing was a priority.

And I felt that a duck pond was a must on any farm, but that was one job that didn't take very long. A wet spot to the front of the proposed site for the house was transformed in a few hours by our good friends the diggers who seemed to thrive on being buried axle deep in a mixture of cow dung and well agitated Welsh mud. And since a duck pond is a bit pointless without a duck, we bought one Khaki Campbell from a local poultry sale, much to the amusement of the big buyers. They'd have laughed even louder if they'd heard what the family called it.

Quackers didn't mind being the only duck, and she took a leaf out of Lammy's book and bedded down with the cattle - although it didn't stop there. Like Lammy again, she was partial to their rolled barley and oats as well. Feed time at the trough was somewhere between a farce and a fiasco, with an orderly line of munching cattle being pushed and harried by a bullying sheep, a fraction of their size but worried in case she didn't get her fair share, and a kamikaze duck running and flapping around the cattle's feet, trying hard to eat as much as everyone else.

The pantomime was completed by the grunts and squeals of an extremely fat pig, standing on her hind legs and looking over the top of the pen door, anxiously reminding us that she had yet to receive her ample rations, and complaining that she always had to wait until last. We bought Annabel as a weaner and her ability to put on weight was phenomenal, partly due to her insatiable appetite and partly to our willingness to satisfy it. Her function in life was to be a breeding sow, although to pass the time during gestation she doubled up as the children's substitute for a riding horse. She was a bit sluggish, and totally the wrong shape, but the only other obvious drawback was a noticeable lack of directional control. Noticeable, that is, by her riders. Wherever Annabel chose to go, the children had no choice but to go too, desperately hanging on and bumping up and down with spine jarring regularity. She didn't object to her secondary function and even seemed to enjoy it, knowing full well that the unaccustomed exercise would work up an increase in appetite and give her legitimate reason to vociferously demand dinner that much sooner.

Thanks to all their help, we were now using quite a bit of 'bought in' feed. Our suppliers, and for just about every farm in the area, was an agricultural co-operative in the Aeron valley seven or eight miles away. The place was a museum, and to walk through the doors of the old wooden shop and store was to go back in time to the end of the last century.

The building was nearing the end of its useful life, but had a character and rural appeal that no modern construction could ever hope to emulate. The shop itself was tiny and decorated with an array of metal advertisements for everything from animal feed to boot polish. The floor was refreshingly cluttered with boxes of various screws, nails, nuts and bolts, and tools of every type and description hung from all available points on the limited wall space.

The rest of the building was taken up with the storage of so many implements, racks and different types of animal feed that I was never convinced that even the manager knew for certain what was there, although, to his eternal credit and our constant amazement, he always seemed to be able to put his fingers on even the most obscure requirement.

He was a small, lean man who bustled around his domain like an overworked broody hen and did everything at maximum speed - unlike his customers whose casual, relaxed attitude more accurately reflected the local way of life. I never once saw him without a cigarette dangling from his mouth or in his hand, and the painful hacking cough that constantly racked his frail body suggested that it wasn't a coincidence.

But he was a gem, a gentle man who could never do enough to help and with a knowledge of available goods that was second to none. And as time went by, and our animal feed requirements increased, we got to know him quite well.

It was encouraging to see our stock numbers grow, but I badly missed not having a dog by my side. It was one of the very few periods of my life without a canine companion. I grew up with a Welsh Collie, a beautiful bitch called Lassie. She started her life belonging to a neighbouring aunt but we quickly became inseparable and the only time she reluctantly wandered home was in answer to an irate call at the end of the day. It wasn't a very good arrangement for either the dog or her owner and to my great joy she was offered to me as a gift. I was only seven, and she was a constant companion and good friend for the next fourteen years. I still think about her from time to time.

My affinity with dogs and other creatures has no beginning other than birth, and I'm one of those peculiar people who believes that all animals are crucial to the mental and physical well-being of the homo sapiens species - and I don't mean on the dinner table. They are as vital a part of the delicate balance between man and Nature as the environment itself, and without them the human race would struggle to survive.

Dogs are in a category all of their own. They have a loyalty to their chosen leader that has no limit, and an unselfish love matched only by a mother for her child. They have compassion and understanding, reacting to their leader's moods with a caring tolerance untainted by personal needs or materialistic intrusion. They have bonded with man yet retained the intuitive instincts of their wild ancestors, an ability they delight in sharing with their chosen 'owner'. Speech becomes unnecessary in the closely psychic and insulated world of a dog and his master.

Dogs are like children, they respond to the love and guidance of those they respect. When denied their deep rooted wish to belong and to please, they can sometimes revert to the drastic survival measures of a lone outcast, banished from the security of the pack. We shouldn't then be in too much of a hurry to blame the dog for subsequent transgressions. Perhaps we should first question the motives and actions of an ill informed 'owner', too busy to give their dog those most precious of gifts - love and time.

There are those who would disagree, but in my opinion a 'superior' species that spends so much of its time and resources inflicting pain and misery on its own kind is not really qualified to pass scant considered judgement on the moral and social standards of other, 'lesser' creatures.

Irene knew of my feelings and for my first birthday in Wales presented me with a pretty little eight week old black and tan Welsh Collie bitch. I named her Pippa. She filled a missing dimension in my personal life and I felt a more complete person having the companionship of a dog again.

Pippa was newly arrived and still little more than a ball of fluff on the day I dug the foundations for the house. Her trust and loyalty were immediate. Curled up on a heap of soil she seemed oblivious to the cold as she watched my every move. Twice I took her indoors in the warm, and twice she followed me out again. It started to snow, and her shape slowly merged with the whitening landscape, but still she wouldn't give in. What a tough, loyal little dog she was, never frightened by anyone or anything, and seldom far from my side for the next fourteen years.

Chapter Six

Domino, the Skewbald Pony / Dusky, the Welsh Mountain Pony

My love of dogs was matched only by Nicky's love of horses, and like most young girls she had longed for a pony of her own from the time she was about five years old. Nothing would have given us greater pleasure than to fulfil her dream, but the right horse would almost certainly cost a lot more than we could afford. It was quietly discussed many times but we didn't dare take a chance and move that far from our projected budget. We promised that as soon as it became possible she would have her wish. Nicky accepted that and didn't mention the subject again, although she continued to jealously guard her modest savings so that one day she would be able to buy one herself.

It was a problem that seemed to have been resolved, at least temporarily, when a local, single lady approached us with what appeared to be a sound idea. For some obscure reason best known to herself she had bought a little twelve hands high skewbald pony, just the right size for young children. It was an illogical purchase to say the least because, as we soon came to realise, she didn't seem to like anyone under the age of about thirty, and certainly wasn't very happy in the company of children.

Free grazing in exchange for as much riding as the children wanted was the proposition. She sold the idea to us well enough to arouse suspicions that she was a part time sales shark, yet we still fell for it. In our defence, such as it is, the proposal sounded plausible. She had a horse that was seldom ridden, guaranteed as safe as houses and totally bomb proof, but very little grass. We had three volunteer riders and plenty of grass, but no horse. It seemed like a good arrangement all round - but then it was probably a good idea to build an unsinkable ship and call it Titanic.

Domino duly arrived, a pretty little pony but with a width that could quickly rupture an unwary rider and with enough fat on his girth to suggest that a lack of grass had not been one of his problems. For the first two hours his head didn't rise above ground level as he dedicated himself to the dual tasks of filling his rapidly swelling belly with the juiciest grass on offer and displaying an aloof indifference to the friendly words of greeting from the children, trotting briskly out of reach every time they walked towards him. Big bare patches in the grass traced his meandering route across the field and his form steadily took on the grotesque proportions of an over inflated rugby ball.

The arrangement was already beginning to look a bit one sided, and there was little doubt in our minds that if Domino wasn't the greediest horse in creation he came a very close second. But there comes a point when even the most gluttonous of creatures has to reluctantly admit that it can eat no more, and as he slowly raised his head for the first time the discomfort caused by stuffing himself beyond the design capacity of his stomach was painfully evident. He was in some distress, and the air reverberated with rasping sounds of gaseous expulsions as he made strenuous and sustained attempts to discharge more body waste than was immediately available in an effort to relieve the internal pressure.

There was a lot more grunting and straining before he finally had to admit that the course of nature can't be hurried, and in the absence of anything better to do while time, and the other more tangible matter, passed, he tottered over to the caravans to see what sort of a home he'd been landed with.

There was still an arrogance about his gait in spite of the self inflicted predicament facing him - or to be precise, facing away from him. People who know nothing about horses will insist that a small pony doesn't have the mental capacity to be patronising, so in that case the shock and dismay on his face as he surveyed our humble abode must have been imagined, a reflection of the reluctance we sometimes felt during vulnerable moments to open our doors for inspection.

Our guest had very quickly shown a natural talent for rubbing people up the wrong way and seemed determined to use his undoubted abilities to their absolute limit and make himself as unpopular as possible. To his credit, he was already enjoying a high degree of success. We were tempted to call it a day and send him back where he belonged, and only a stubborn refusal to be intimidated and beaten by a jumped up runt of a pony with no form and dubious parentage made us keep on trying.

The cause of peace and co-operation can sometimes be helped by a bit of bribery, and Irene thought she'd pour oil on troubled waters and try to win him over with a carrot. He'd obviously tasted them before and his eyes lit up like beacons. Ignoring the fact that he was already full to bursting point, he greedily thrust his head forward to grab it before she changed her mind.

Now it can't be denied that through a horse's eyes a finger probably looks very much like a carrot, and the rest of the family was prepared to give him the benefit of the doubt and concede that a mistake had been made. However, the anguished yell that followed his unfortunate error confirmed that Irene did not agree, and that she had no doubts at all about his intention. I regret to say that she did not react to the pain in her already swelling digit with the degree of understanding and forgiveness that would have undoubtedly been displayed by the rest of us and applauded by the urban horsey brigade. And worse, her displeasure was physically reinforced with a hefty and accurately delivered right hander across the end of the pony's nose that made him gasp for breath and brought tears to his eyes. It was a reaction he'd never met before and didn't much like.

Retaliation was inevitable and his portly form was galvanised into defensive action with a surprisingly agile jump through a hundred and eighty degrees that left his backside glaring menacingly towards her. The sudden bodily movement had the secondary effect of stimulating his disposal system into spontaneous action, discharging the long awaited waste matter with explosive force to land meaningfully just six inches from Irene's feet. It was a point well made and easily understood.

Their friendship was thus remarkably short lived, and the prospect of reconciliation looked about as likely as flying pigs. While Irene nursed a bruised and bleeding finger, a slimmer and completely unrepentant pony triumphantly retraced his steps, forcefully helped on his way by a deluge of choice words that could only have been learned from Megan. He had doubtless heard them all before, and casually turned his attention once more to grazing and to the less entertaining but more rewarding business of refilling the unexpectedly available space in his ample paunch.

The thought of actually riding this obnoxious little beast had suddenly filled our budding jockeys with no small amount of trembling appre hension, and in line with one of their rare unanimous agreements it was decided to postpone the dreaded event until the following day. The twenty four hour reprieve presented a golden opportunity for a little bit of youthful scheming, and it was decided by a majority of two to one that Paul was best suited to the task by reason of his skill and courage, and being ganged up on by his brother and sister. Never one to duck a challenge, he faced his moment of truth with a show of stoicism and nonchalance as far removed from the fluttering sensation in his churning stomach as bravery is from cowardice.

Domino stood quietly and obediently as the bridle and saddle were positioned and secured, innocently staring into space without a disruptive thought in his head. Paul prepared to mount, confidence beginning to course through his body in the wake of an adrenaline overdose. His first big mistake was to assume that he really was on a bomb proof pony, and to settle himself into the saddle with a slack rein. That was a sign to Domino that his rider was at his most vulnerable, and he took off across the field like an express train, surrounded by a cloud of condensed breath from horse and rider and followed by flying divots of good Welsh soil. His second mistake was not firmly locating his feet in the stirrups before settling himself in the saddle. The resulting brief contact of his posterior with the horse's back became more and more spasmodic and his rise above it became greater and greater with each bounce. Viewed from the rear, the temptation to compare the phenomenon with a Thelwell cartoon was overwhelming. Paul hung on grimly like a true champion, but an excursion into the sphere of aerobatics was inevitable. At the extremity of the field, just as he was bracing himself for a skidding stop, Domino changed course. Paul didn't.

That was the last straw really. We decided that other people's bomb proof ponies should stay with other people in the future. We'd manage in our own way in our own time, as we always had in the past.

Another of our neighbours had a few horses as well. She was a local character who was well known for a variety of different reasons by just about everybody for miles around. She had been widowed some years earlier and now worked all hours of the day and night trying to run a pedigree herd of Jersey milking cows on her own. What she didn't know about the Jersey breed wasn't worth knowing and her knowledge and experience were regularly utilised by the organisers of agricultural shows looking for credible judges.

Unfortunately Sally also had to adhere to the rules and demands of commercial logic, and that was where she unfailingly came unstuck. Remaining solvent had been an uphill battle for as far back as she cared to remember, and paying her bills was an unfamiliar task that she consistently tried to put off for as long as possible, in spite of a daily deluge of requests for her attention to long outstanding debts. She spent her life ignoring anything remotely conventional and lurching from one crisis to another. At times of extreme pressure from either her suppliers or her bank manager, which were frequent, she'd sell anything available to get herself out of trouble, except of course her beloved cows. She'd reached the point where there wasn't that much left to sell. The farmhouse had gone years before and her only remaining home was a tiny store room leading off from the cow shed, which served as bedroom and kitchen and was prone to periodic attack by a succession of hungry rats wanting to know why she didn't keep the cow feed there any more.

Our daily visits to buy her rich, creamy Jersey milk inevitably included an update in the apparently endless saga of her downward plunging spiral, and it came as no great surprise to learn one day that her month's milk cheque wouldn't be enough to stave off yet another threatened recovery action. Something else would have to go. This time apparently it was the turn of the little Welsh Mountain colt. He was still with his mother in the same field where he'd first seen the light of day a year before. He had good breeding but he'd never been handled and was as wild as they come.

The young pony represented temporary salvation for Sally. He would provide her with another short period of respite and allow her to rebuild her strength and resolve before the next emergency erupted with its unfailing predictability. She ought to have been happy at the prospect but we found her in sombre mood. There was only one way to sell the pony quickly, and that was through the auction ring, but with no control over the identity of the buyer or his purpose in buying, Sally knew that he could face a very uncertain future in Belgium or France. She found the prospect difficult to accept.

The solution to everyone's problems was obvious and Nicky didn't need to be asked twice. Parental caution was dismissed with one of those beaming smiles endowed by Nature on all young girls, complete with a cast iron guarantee that they'll melt any father into immediate and unconditional submission.

The necessary formalities were completed within hours, and the following day the whole family embarked on a mission to collect Nicky's new purchase. Taking him away from his mother was going to be an exercise unlikely to receive his wholehearted approval, so it was suggested that since Sally knew her horses better than anyone she was entitled to plan an appropriate strategy. It was a proposal based on observation and blatant cowardice. We'd only seen the pony a couple of times and on each occasion he had seemed equally at home either standing on two legs whilst knocking hell out of an unseen opponent with the other two, or galloping around the field at an unstoppable pace.

We needn't have worried because Sally's plan was foolproof. She would lead the mare, loyally followed by the unsuspecting offspring, into the yard, at which point they would be separated and a head collar quickly strapped onto the colt's head. The addition of a couple of lead reins would complete the capture. Nothing could be more simple, and from a safe distance we watched in silent admiration as the first part of the operation worked without a hitch.

Now it has to be said that Sally was not a small lady, and years on the farm had given her muscles that would have been the envy of many a young rugby player. So it was with supreme confidence that both arms were wrapped around the pony's neck and a head collar drawn slowly towards his nose. It was at this point that the plan showed itself to be less than perfect and our confidence nose-dived. Without warning she found herself lifted clear of the ground with a speed and athleticism that she hadn't known for forty years, and her considerable bulk was swinging in every conceivable direction like a demented pendulum, her feet occasionally dragging across the rough concrete floor before taking off again on an aerial route around the yard.

There was very little doubt in our minds that unless the pony became giddy and gave up, which in view of his daily circuit training in the field was not very likely, Sally was not going to win the battle. The pony had arrived at the same conclusion. He quickened the pace of his spin until, with a well timed toss of his head, Sally's hold was broken and she landed flat on her back in the middle of the muck heap, arms outstretched and green welly boots waving frantically in the air. The smell as she staggered to her feet was awful, but it gave legitimate excuse for the coughs and snorts of onlookers as hysterical laughter threatened to erupt through quivering lips, which no doubt would only have added to the uncharacteristic expletives emanating from the direction of the heaving stench.

Our concerned enquiries regarding her current state of health were answered with a cold stare, just visible through strands of hair firmly stuck together with recently discharged excrement. The delicate situation wasn't helped when one of our number, who shall be nameless, finally lost control and got a fit of the giggles, which did nothing to help Sally salvage what little remained of her feminine dignity.

The operation was eventually brought to a satisfactory conclusion by reinforcements drafted in from neighbouring farms, headed by John who had agreed to face daylight and leave his farm only on the promise of a one to one wrestling contest with the equine transgressor. By this time any regrets Sally may have felt at parting with her pony were well and truly extinguished and she departed without a backward glance for a hose down in the cow shed.

John's mission was now completed and with the smug look of satisfaction that goes with a job well done, he left for home and we were on our own, except for Dai and Megan's daughter Rhian, who had stayed on in the hope of a bit more live entertainment. The pony wasn't impressed by his new opponents and appeared to rate his chances somewhere between very good and excellent. Once again, his assessment was fairly accurate. Having beaten one leader into submission he now viewed his audience with utter contempt, daring another volunteer to try his luck. The look in his eyes suggested that negotiation was out of the question.

With thoughts still very much on Sally's undignified exit, the wisdom of age decreed that the answer to the whole problem was to be found in the agility of youth, and Nicky, her two brothers and Rhian were charged with the task of walking the pony a mile or so along the narrow country lane to his new home.

And so another foolproof plan was composed. The two boys took up positions at each side of the pony's head, each securely holding one of the lead reins, and the girls stood at the rear each holding one end of a strap held tight against his hind quarters. The pony was motionless for the first time that day. He even seemed to be enjoying the experience, but we put that down to his low cunning and an attempt to lull us into a false sense of security.

I banished a wave of sympathy for the tethered animal that was threatening to flash across my mind, and gave the order to move out in a manner that seemed appropriate at the time but which only succeeded in prompting more giggles. In their own time the handlers heaved together and the pony walked quietly and sedately between them towards the road. It was working perfectly.

After twenty yards he had a change of heart. He stopped, dug all four feet into the tarmac surface and refused to budge. Threats, pleas and much waving of arms failed to elicit even the smallest degree of co-operation, and the operation was called off to give serious thought to forming an alternative plan. That convinced the pony that he had won another round, and he took off down the road dragging two panting boys on the end of lead reins and with two young ladies sprinting along behind. After fifty yards he changed tactics again and stopped, dug his feet in and refused to move.

Two things had already become abundantly clear. First, it was apparent that whilst the chosen strategy was without doubt working, it was more than likely that, if the current rate of progress was maintained, it would take the rest of the day to reach our objective. Second, the sound parental advice that had been periodically offered as Irene and I wandered leisurely down the road behind the sweating children had not been heeded, or indeed appreciated.

Coincidentally, it was at this point that I remembered that all children need to know that their parents trust them, and now seemed like an excellent opportunity to display a belief in their abilities and go home for a cup of tea.

An hour later a quintet of steaming bodies appeared at the farm gate and, still progressing in fifty yard bursts, gratefully negotiated the last lap down the track towards a freshly prepared loose box in the cow shed.

Once inside all hell broke loose, but the pony soon discovered that testing the construction of his new home by systematically kicking each wall in turn caused more discomfort to his feet than to the building, and he began to settle down and look more relaxed. His relieved and exhausted conquerors collapsed onto bales of hay and triumphantly relived every minor detail of their epic journey.

The task of giving him a name was a little less harrowing. He'd been called quite a few things during the course of the day but none suitable for a young girl to use. We decided not to ask Sally for her suggestions. It was Nicky's horse and it had to be her choice. You would expect a horse's name to reflect his character or personality, so the reason for her final decision was a mystery that was never questioned or explained. She may have been influenced by the aristocratic names of his parents, but more likely it was because of his colour. But for whatever reason he finally became known by the poetic name of Silver Dusk.

And so "Dusky" came into our lives, and a much loved daughter was very happy.

Chapter Seven

Building the House / The Village Gossip / Ali the Lamb / Holy Grail / Caravan Awning / More Dogs

As the house grew and took on its final shape there was a lot of interest by passing local people, and a few called in on some pretext or other to have a closer look.

The village gossip couldn't take her eyes off it. She was an elderly farmer's wife, a diminutive woman with pointed features and a macho beard. On most days of the week her clothing betrayed her way of life, a lace edged blouse over a drab tweed skirt carefully selected to cover every part of her modest frame, and extending all the way down to a pair of hard wearing and sensible country brogues. Sundays were different. On that special day she could be seen walking at her usual brisk pace towards the chapel, colourfully attired in one of her best floral patterned dresses that dared to show an inch or two of calf, and topped by the most frivolous of wide brimmed hats with an extravagant feather drooping forlornly down one side.

Irrespective of the day of the week, her tongue maintained a razor sharp edge that was frequently brought into action, and she ruled her household and even smaller spouse with an amazonian fervour. Yet when anyone was in trouble she would be first on the scene, complete with a heart of pure gold.

She liked to know everything about everyone and looked on newcomers as legitimate fodder for local gossip, acting in the firm and genuine belief that members of the community who had been born and bred in the vicinity had a right to know about all the various goings on. There was nothing personal about it, as a responsible village elder it was her duty to seek out and to pass on any juicy information.

It was fair to say that she enjoyed her work. Every single day without exception she drove up and down the lane to make sure she hadn't missed any minor details or recent changes that should be reported back to the community at the earliest possible opportunity. She took her duty very seriously and gave it maximum attention, sometimes more than was wise. Every now and again her mind wandered too far from the Highway Code and the little car would wobble dangerously from one side of the narrow country lane to the other.

She misjudged it one morning and car and driver came to an abrupt halt, nose first into the roadside ditch. There wasn't really anything funny about it, after all she could have been hurt, or at the very least badly shaken. But the sight of the car's rear end pointing skywards with its composed driver striding manfully and gleefully down the driveway, finally armed with the legitimate excuse she'd longed for, brought out the worst in us and one or two smiling faces had to be tactfully turned in the opposite direction.

So having watched us put her car back where cars belong, she was given a guided tour of the holding and satisfied weeks of pent up curiosity by looking over the inside of the house and outbuildings, and generally weighing up everything else in sight. The car was undamaged and all in all it had been a good day for her, and no doubt it became even better as she transmitted everything she'd learned to anyone interested enough to listen.

We did wonder if the whole thing had been planned, but if it was we weren't treated to a repeat performance, although her erratic course down the lane continued daily until there was nothing new to see.

It was fortunate that she didn't see Ali during her tour, otherwise the relayed report to her friends would have taken on an added dimension, and our new found reputation would have tumbled back down to rock bottom again.

Ali was a twin lamb reared from one of our Welsh Mountain ewes. Her mother had unilaterally decided, as sheep sometimes do for a variety of different reasons, that this particular offspring was nothing to do with her, and one was enough to take care of anyway. So, like Lammy before her, we became her substitute mothers and she was reared on the bottle. But unlike Lammy, Ali wasn't satisfied with a straw bed in the cow shed and she wasn't interested in calves either. Her ambitions went far beyond that.

Her full name was Ali Baba, named by the children for obvious corny reasons, and she followed them everywhere, including into the unfinished house where they'd taken up daytime residence and put up an old discarded camp bed for her in one of their future bedrooms.

We didn't mind at that time, and let them enjoy their bit of fun. The problems started when we finally succumbed to the growing temptation to move into the house ourselves and Ali's eviction became necessary. It wasn't as simple to get her out as you might think. She'd got used to home comforts and wasn't going to give them up without a fight. Whenever the door was left open, in she'd go, straight up the stairs to bed.

It was a problem that took time and patience to resolve, but not before it had left an indelible mark on one or two other people as well.

Close family members were spending a few days with us, and they'd parked their touring caravans in fields to the front of the house. It was an idyllic position and they were happy to just soak up the scenery and enjoy the peace and quiet, including an ageing uncle who had some time ago reached the end of his working life, and as a self prescribed means of keeping body and soul together was known to indulge in a little well earned medicinal scotch now and again. As we heard the story, Ali was responsible for him coming dangerously close to going on the wagon.

He was engaged in the popular holiday pastime of doing nothing with enthusiastic vigour, reclining in a framed canvas chair outside of the caravan, a half empty glass in his hand and staring vaguely across the duck pond to the hills beyond, when he glanced casually towards the house. He knew it wasn't possible and that everyone would laugh at him, but he was convinced that a sheep was watching him from the landing window.

Summoning up all his courage, and already doubting his sanity, he called for my brother in law to confirm what he thought he'd seen. By that time only the vague outline of curtains was visible behind an otherwise empty window.

Family reaction was predictable, and the poor man suffered hours of good natured ribbing about the unfortunate effects of combining nearly eighty years of life with best scotch whisky.

The truth was only known to him the following day when my brother in law's usual cheery greeting round our half open front door was answered by the sound of feet landing heavily on uncovered floorboards upstairs. As he watched, Ali appeared on the landing and, thinking she was being called for supper, clip clopped down the stairs with a nonchalance befitting her claimed status of a family member.

Ray's disbelieving laughter came close to the point of hysteria, and the creation of an accidental puddle on our bare concrete floor would have come as no great surprise to any of us.

The mental well-being of that same uncle came under suspicion again a few days later, only this time he had a witness, or possibly another contender for certification. At the rear of the house was a 'hog back', a small hillock gently rising some fifty to a hundred feet before falling away at the same angle on the other side. He and my sister had wandered to the top for no reason than that it was a pleasant spot that provided an excellent view of the hills and the sea.

It was a bright sunny day, and as they clambered over the ancient stone wall that ran along its ridge, the shape of an enormous Christian cross in the grass was made plainly visible by changes in the shades of green. They confirmed to one another that it was really there then hurried back down the hill to call others to witness this strange phenomenon. Needless to say, by the time we'd reached the spot indicated there was nothing to be seen except grass - all the same colour.

So was it advancing senility and imagination, as we insisted at the time, or an innocently witnessed optical illusion? Or perhaps they were lucky enough to be in the right place at the right time, when the length of grass and the angle of sunlight combined to resurrect an ancient mark. We'll never know for sure but the area has an interesting history that may be relevant.

Our land was on the ancient route between Aberaeron harbour and Strata Florida abbey high up in the mountains, an abbey with strong historic links with Glastonbury. Legend has it that the Holy Grail was brought from the Holy Land by Saint Augustine and delivered to Glastonbury. There it rested, undisturbed until the abbey itself was subjected to the paranoid dissolution policy of one of our less appealing monarchs. It was then, the story goes, taken to Strata Florida.

If the story has any truth there would have been many subsequent pilgrims arriving by sea and they would almost certainly have passed through or by our land. It's equally certain that one or two travellers of advanced years would have lacked the required stamina and expired en route. Hence the cross on the top of the hillock?

The story doesn't end there. In due course Strata Florida Abbey itself was put to the torch, but not before the Holy Relic had again been moved to a place of safety, this time to Nanteos House, a family home of some substance near Aberystwyth. As the family dispersed and the estate subsequently fragmented, the details of its fate became a little hazy, but there are a few people, including at least one clergyman, who claim to know of its current whereabouts and have provided film evidence in support. No doubt some sort of relic does exist in West Wales but its authenticity has never been proven and, like the Holy Shroud of Turin, is probably of doubtful age and origin.

There are many legends surrounding the Holy Grail, and this is probably just another fairy-tale for those who want to believe it, a story for the gullible, or even a commercial ploy by the monks. But it's also remotely possible that it contains a grain of truth. It may be speculation to a point of absurdity, but no one has yet come up with a better explanation for the existence of the cross at Mesen Fach.

Surprisingly perhaps, our family came back again and stayed with us many more times. Usually it was for a quick break, but on one occasion it was also their base for an additional week spent on a canal boat in Mid Wales. The caravan wasn't very convenient for that length of time, so they'd decided to spread themselves out a bit and had bought a smart new awning, suitably furnished with a table and chairs and all the other various odds and ends that go to make caravanning that important bit more comfortable and homely.

During their absence we made one of our routine visits to the local mart, leaving the place empty apart from a few cattle grazing happily in the field around the caravan.

It was a beautiful day, and as we arrived home a slight breeze was blowing up from Cardigan Bay, gently rustling the beech leaves in the long hedgerow bordering our land. Our few trees were lazily waving in friendly welcome and Pippa voiced her pleasure that we'd come safely back to her. Even the new awning was flapping noisily as though in greeting.

The pleasure of being home again drove minor problems from our minds and we felt relaxed and at ease with the world. The peace and tranquillity somehow found its way into every part of our beings and we were struck with the realisation that, for all its warts and unfinished chaos, it was a beautiful place and it was our home.

We were out of the car and indoors before reality thrust its ugly presence back into our minds. The awning shouldn't be flapping. A terrible dark foreboding descended on us and we rushed back up the driveway, desperately hoping that our worst fears wouldn't be realised - alas in vain.

The entire side of this beautiful new expensive awning was ripped apart from top to bottom, its contents exposed to the world and framed by the ragged edges of the now useless canvas.

We didn't have to search far to find the culprit, and we weren't fooled by the look of childish innocence as we approached her. There, still skewered on the end of Funny Girl's right horn, was a telltale remnant of canvas awning.

When confession time arrived we humbly told the story with straight faces and sincere contrition - until I handed over the recovered remnant. Then for some reason hysteria took over.

Animals are notorious for getting themselves, and their owners, into trouble, and the Mesen Fach stock seemed to be working hard to maintain that reputation, and destroy ours in the process. Not content with just witnessing the disruptive talents of our cattle, Pippa decided to join in the fun and make her contribution to the daily bedlam that is called stock rearing by laymen and masochism by the initiated.

One of our immediate neighbours had a similar dog, and they each protected their own land so didn't get on too well. It was all threat and no action, but with plenty of verbal sparring. Each time they saw one another through the boundary fence we were treated to the mandatory sabre rattling display of courageous growling, snarling and rushing ferociously up and down the fence line, followed by the sort of crouching glare that dares the other to make the first move.

We're not entirely sure who did eventually take the initiative, but unknown to us at the time there was apparently at least one occasion when a temporary truce was negotiated and they became very close friends. The result was Trixie, Tina and Candy.

Trixie was the most beautifully marked of them all, and her long shiny coat glowed with vitality and good health. She was a happy dog, transparently honest and as agile as a cat, but a little unsure of herself. She wasn't the most naturally gifted of collies but she'd work non stop until the job was done, which more than made up for any shortcomings. Her only aim in life was to please us, and most of the time she succeeded.

If there was a negative side to Trixie, and you would have to look hard to find one, it was her habit of leaping into our arms for a quick cuddle and a little bit of reassurance. None of the other dogs could jump that high and she would smile down at them with a contented look of arrogant superiority. To be honest we didn't discourage her, it was a bit of fun that gave her the confidence she lacked even though it could be embarrassing at times. She didn't always choose the right moment and leaping into my arms when we were chatting to visitors was bad enough, but timing her jump to coincide with a mouse running past my feet was, to say the least, unfortunate. Their conclusions were obvious and poor Trixie never lived it down.

Tina had the best sheepdog's instincts by far, and before she was weaned would practice for hours, grouping all the chicken together or moving poor old Cocky round and round the yard until he couldn't take any more and ran off down the driveway flapping his wings at maximum speed before taking off and skimming the top of the adjoining fence to clumsily crash land in the adjacent field, still squawking loudly in hopeless defiance.

She would sit watching penned sheep to make sure they stayed put, and if one dared to move she prowled up and down the pen, head close to the ground in the classic hunting position, until she was satisfied that her point had been made and understood. Unfortunately, as she grew older, her attention span in the fields proved to be short, and when she felt that she'd done enough she simply wouldn't work any more, which tended to prompt the sort of language not usually associated with the training of sheepdogs. All this pleased Trixie, who would laugh in the way that dogs do, finish the job on her own and wait expectantly for the inevitable verbal applause and fuss. The rivalry between them was intense.

Candy was the odd one out, jet black with a white blaze that started high on her head and ended behind her front legs, which is why she was affectionately called badger bonce. She had the most appealing and expressive brown eyes you will ever see, and could get away with any misdemeanour simply by looking upwards with an innocence that only the hardest could doubt.

She was at least as intelligent as her sisters but it suited her best to pretend to be stupid. She was clever enough to get away with being lazy, and was never known to bury her bones as instinct told her to do. But then, she operated a system that made it unnecessary. Using her inherited sheepdog cunning, which she successfully avoided using on the real thing, she would crouch behind any handy cover and watch where her sisters buried theirs, then wait long enough for the coast to clear before digging them up again to add to an ever growing store in her own private larder.

Candy loved eggs as well, chicken's or duck's, she wasn't fussy about their origin, and it was always a race to find a new nest in the hay. Candy usually won, and it was a common occurrence to see her quietly sneaking off to some secluded hideaway, a new laid egg gently held in her soft mouth, and furtively glancing from side to side in case someone was watching, to feast on her ill gotten gains. Small wonder she had a tendency to put on weight.

As a farm dog, Candy was less than useless, but we all loved her just the same.

Chapter Eight

In the House / Home Baked Bread / Dozy the Cow / Dusky Again / Tammy the Arab

Since our arrival we'd managed, to some extent, to achieve a modest level of self sufficiency. With guidance from Megan, Irene had become a proficient baker, and the more usual mass produced, chewy concoction that passes for bread in most shops these days had quickly become a thing of the past. Three or four times a week a fresh batch of home baked goodness came out of the oven, filling the house with a succulent smell that made mouths water and brought three children rushing to the kitchen for the first hot slices thickly covered in melting butter.

The bottomless pits that they called digestive systems were enough to guarantee that none of it lasted long enough to go stale, and the butter store had to be regularly replenished. The latest results of Irene's new found expertise were always awaited with eager anticipation and impatient prompting. My part in the procedure was to make sure that there was always a good supply of flour for the next batch, bought from our animal feed suppliers in hundredweight bags and stored in a milk churn to keep it dry and fresh.

We were still buying our Jersey milk in gallon containers from Sally, which in most respects was a good arrangement, but what in theory should have been a simple collecting job had developed into a time consuming event that on occasions had drifted on for a couple of hours. Sally always welcomed an opportunity for a chat about her cows, and we became familiar with the breeding of her prize winning specimens right back to the year dot in the Channel Islands. It was all very interesting for those who had chosen to give their money and their lives to the cause, but we weren't among them and had other things in our sights.

More often than not, she prolonged the visit even more by throwing all the depressing details of her most recent financial crisis into the one sided conversation, which left the listener with nothing positive to say in response, except perhaps a few choice words of advice that she wouldn't have liked, had probably heard from other sources many times before, and would have ignored anyway. Our reluctant silence tended to be punctuated only by the occasional tacit agreement, which made her feel better and got us off the hook.

If we could get her away from those two subjects she was a very interesting lady with a colourful past and a family history which surprisingly included a pampered upbringing by her high ranking military father and well bred mother - as far removed from her present existence as chalk is from cheese. But since those stimulating occasions were becoming increasingly rare to the point of imminent extinction, it had become a daily routine we could well manage without. In any case we really needed to be more independent and to have our own dairy cow.

We'd mentioned to the village gossip that we were considering buying one, and it just so happened, coincidentally of course, that she had a quiet and freshly calved young cross bred for sale. We were on the point of buying it too, until another neighbour tipped us off that it had recently gone down with milk fever, a condition that, depending on your point of view and the level of calcium in the soil, can possibly recur at each calving.

It was a problem we'd be likely to face again. Country people are as honest as the day is long, although when livestock is changing hands the day can sometimes shrink a little. The unwritten laws of the market place demand that all faults are disclosed, but when a private purchase is being negotiated the gloves are off and it's wise to ask yourself why the animal is being sold and to look very closely to see what's wrong with it.

We really needed to buy from a herd we were already familiar with, and that gave us a choice of only two. We knew most of Sally's cows quite well after doing some relief milking for her on one or two occasions, and when you're used to thick, creamy Jersey milk anything else seems little better than mildly flavoured water.

I must confess to a liking for rich dairy products. The attendant health risks pompously proclaimed by clever dicks masquerading as nutrition 'experts' (whatever that word means) have the same effect on me as water on the proverbial duck's back. They may have some relevance when applied to overweight pen pushers who spend their entire lives on office chairs during the day and armchairs in the evening, but for the rest of us who use food as a fuel to be burnt up during a day's labour, it's one of the many misleading examples of a society that has chosen to reject a more natural way of life. How's that for tempting fate?

The other deciding factor was Sally's ongoing need for a capital injection. Her level of liabilities had reached crisis proportions again, and it seemed like a good time all round to ask if she had a heifer to sell. We were pleased, but not too surprised, when she grabbed the opportunity to pay some of her bills for that month with both hands.

And so we became the proud owners of a pretty little pedigree Jersey heifer, freshly calved, full of milk and as quiet as a lamb. Quite a bit quieter than average in fact. Cows aren't noted for their intelligence, but this one was as thick as two short planks. But then, we should have known that when Sally told us her name.

It was a very minor problem and Dozy was good to us, giving us a gallon of milk every morning and with two or three gallons left over through the rest of the day to feed to the calves. Each evening, after cooling in the fridge all day, there was more cream skimmed from the top than we could possibly eat ourselves, and the dogs fed like royalty.

Sally never referred to her embarrassment on the day we collected Dusky, and we assumed that she wanted to forget the occasion - and the pony. We did just casually mention that he'd settled in very well, but the information fell on deaf ears so we left it at that. In fact he had very quickly worked out that a clean dry stable every night and as much grass as he could eat during the day combined to give him a luxurious lifestyle that was infinitely preferable to his last home, however loving and accommodating his mother and owner may have been, and had unilaterally decided to become a permanent fixture in our ever growing family. Nicky had more than lived up to her promises and with the help of her two brothers had spent every spare hour, first halter training and then teaching him to accept a saddle and bridle. He seemed to enjoy the experience, and before too long had responded to the mountains of love lavished on him by a devoted owner and they had become good friends.

From the very beginning it had been obvious that Dusky was an extremely intelligent horse. He always insisted on being the centre of attraction, and given the faintest hint that he was being ignored would quickly think up some new stunt to grab our attention. His sense of mischief became legendary, and he had an uncanny knack of winding people up until they twanged. The resulting shouts of anguish, and the occasional smack across the nose were, to him, moments of supreme pleasure and triumph. We got used to him and accepted his antics as a normal part of everyday life, but visitors were usually taken a bit by surprise.

There was one unfortunate electricity meter reader that we only ever saw the once. The meter was high on a wall in the barn - Dusky's territory. We did warn him, but he assured us with a supreme confidence and a slightly superior smile that he'd been brought up with horses and there was absolutely no need to be concerned. So, having done our duty and cleared our consciences, we stood back to see what would happen, knowing from bitter experience that the next few minutes would be anything but boring.

Dusky didn't disappoint us. Just as the poor misguided man was straining upwards to get a good view of the relevant figures, he found himself lifted clear of the ground, his feet kicking wildly and vainly in a haphazard sequence of circular and vertical directions until his new found friend decided that the game was over and unceremoniously dumped him back onto the hard concrete floor.

We suggested that Dusky was probably only helping him to see the figures more clearly, which was obvious to us but for some reason he couldn't find it in himself to agree.

No one was safe, as sister Sybil discovered the hard way. She and Ray were spending a few relaxing days with us again, and she was wandering aimlessly across the fields without a care in the world. Dusky was grazing happily nearby and didn't appear too interested in anything except his second favourite pastime of filling his stomach. Thus Sybil became yet another trusting soul lulled into a sense of false security, her mind drifting away onto vague and irrelevant issues far removed from intransigent horses. That is, until she realised that Dusky was following her.

ELECTRIC

As she quickened her pace, so did Dusky, and as she moved to one side he did the same, and all the time he was getting closer. Before she knew what was happening, she was as near full gallop as she could manage, and looking over her shoulder more often than was wise. By the time she realised that the duck pond was her only means of escape it was too late to change course.

We'll never know what would have happened if Paul hadn't chosen that moment to look out of the front door. Sybil was certainly pleased to see him, but I can't help feeling that there would have been a much more interesting end to this story if he'd stayed indoors.

As it was, faced with charging masculinity, Dusky retreated. Playtime was over and he returned to his meal.

Dusky's future was assured, but there was one horse in the nearby village that, according to local gossip, wasn't quite so lucky. A beautiful little Arab gelding had apparently been abandoned in a small paddock about a mile from our home. The adjoining cottage was suddenly empty and deserted and no one knew where the owners had gone or why. It was a situation guaranteed to prompt hours of discussion over boundary fences, and provide fuel for numerous speculative tongues.

Everyone agreed that something had to be done, the expected change in the weather was almost on us and wind chill would soon become a ruthless enemy that the outcast would almost certainly be too weak to resist. He was already huddled close to the meagre protection offered by leafless trees and spindly hedges, and his lean and athletic frame was dismally failing to provide the resistance offered by his more hardy, indigenous cousins. The pathetically short grass was rapidly disappearing in a sea of mud, made worse by routine excursions to salvage what little food remained, and he was growing progressively more hungry by the day.

A lady in the village took the initiative by offering a spare loose box and winter feed, and in the space of a single day the pony's life changed dramatically. In the warm dry stable that was now his home, misery became a distant memory blurred by a contentment that he didn't understand but wouldn't question.

The owners were never seen again, but nobody cared. The problem was resolved and we forgot all about it. We didn't dream at the time that during the following few months we would come to know the pony very well indeed.

It was during this period that we acquired another daughter. Not literally you understand, natural additions to our family definitely did not form part of the plan, and in any case our living conditions had not been conducive to the creation of babies. No, this was a young lady from a neighbouring village who had teamed up with Nicky on her first day at the new school. Annabel had become a frequent visitor, and was more likely to find a reason to stay overnight than she was to look for an excuse to leave. She lived with her divorced mother, a woman dedicated to a number of diverse careers each of which contributed to a greater or lesser extent to the domestic budget. Time with her daughter was therefore a luxury always likely to be in short supply, and when she sold her house and moved into temporary accommodation, a lack of available space for both of them cut it back to zero. So it was arranged that Annabel should move in with us until her mother could bring some sort of order back into her life again.

The common bond between the two girls was a love of horses, and their spare time together was inevitably spent riding. Nicky and Dusky had become inseparable, but at that time Annabel didn't have a riding horse of her own so she borrowed the abandoned pony in the village, using the plausible excuse that he needed to be exercised. His rescuer was working wonders with him, but regrettably she had indicated a marked lack of either good eyesight or adequate observation by cursing the unfortunate animal with the gender bender name of Tammy. Standing back to view his athletic frame, I failed to understand how she could possibly have overlooked the obvious appendages of a gelding. However, he seemed to be oblivious to any implied slur on what was left of his manhood and was becoming a picture of good health and happiness, very spirited and 'a bit of a handful'.

By this time Dusky had also been the subject of veterinary interference with his breeding apparatus, so they already had something in common - or didn't have, depending on which way you looked at it. The surgical snip had been an unfortunate but necessary intrusion into his private life. Stallions are notorious for ignoring the wishes of their riders on each occasion that some doe eyed mare living within a two mile radius has reached that period in the year when her hormonal request for male company is carried obligingly on the wind to every willing participant in a little equine hanky panky.

Safe from the distractions of such impure thoughts, the two greys enjoyed the company of their own kind and developed the same firm friendship that was shared by their riders. The girls spent many happy months riding them across the fields and along quiet country tracks and lanes before fate decided that it was all too cosy to be allowed to continue, and threw a couple of hefty spanners into the works.

Annabel's father ran a trekking centre high up in the mountains and, having noted that his daughter was showing considerable promise as a show jumper and point to point rider, came to the conclusion that she should have a horse of her own. It was a gesture of parental love that would normally be difficult to fault, but Tammy had become used to being ecstatically happy for the first time in his life and as a result of Annabel's good fortune he would undoubtedly become surplus to requirements again.

His rescuer chose that moment to make her annual attempt to balance the books, never a pleasant task, which in this case aggravated the situation even more by clearly showing that the cost of keeping him in the manner to which he had only just become accustomed was more than she could afford. Even worse, she would have to claim ownership and sell him.

It was all a bit worrying. There was really only one place where he would immediately be relaxed and happy, and it didn't need a lengthy discussion to tell us where that place was. We all do unwise things in life, especially those of us who are prone to an hereditary softening of the heart when children or animals are in need. Put the two together and you have a lethal combination that's almost impossible to ignore. The asking price was modest, but not modest enough, and it took three generations of our family to come up with the solution. And so, displaying the same illogical reasoning that had become our trademark, the purchase was completed and Tammy became Nicky's second horse.

The two horses were total opposites in breeding, size and temperament, but they quickly settled down to living together, sharing the same field by day and the same stable at night. During the school term they lived a life of idle luxury, reclining lazily in the sun to recover from an over indulgence in the ample grass and, when the mood took them, challenging one another to a race across the fields, always knowing that a Welsh Mountain had no chance of beating an Arab.

Tammy learned that life with Dusky could be fun. Most of the time he reacted to the teasing antics of his extrovert friend with the patient tolerance of an understanding mother, but now and again Dusky would find himself on the receiving end of a bite on the flank or a kick in the ribs as a sharp reminder that he had gone too far this time. An unshakeable bond developed from a solid foundation of mutual respect and understanding.

We'd come to Wales with one pet rabbit, and in less than four years our family of pets had grown to two horses, four dogs and a cat that had followed Pippa's lead and multiplied to four as well. Displaying a lack of wisdom had become a habit.

Chapter Nine

Work Completed / Another Farm / Storms

By the summer of 1981 we were well established in the house and with more comfort than we'd known for years. The mains electricity had been connected, and our generator, having served us so well, was relegated to stand by status again. We'd reluctantly renewed contact with the world beyond our small village and could now be contacted by telephone. And as an unexpected bonus, the local water company wanted permission to lay a new service main across our land and had agreed to connect us to it as part of the deal.

It was the extra space after the cramped conditions in the caravans that gave us most satisfaction, although we couldn't deny a certain childish pleasure at experiencing modern conveniences again, simple things like turning on a tap to fill a kettle and flicking on a light switch as we walked into a room - without starting the generator first. We'd got used to being without those luxuries and whilst we didn't miss them too much at the time, it was still good to have them back again. Having central heating and constant hot water was self indulgent icing on an already enjoyable cake.

So building work was finished, the land brought back to good heart, fenced and stocked to the limit. The holding had reached the peak of its potential market value, and any more development or improvements would be money down the drain. The first part of our plan was completed.

It was satisfying but sad at the same time. We'd grown to love the place, and in spite of, or perhaps because of, the hard work and material deprivation, the children had developed into people that any parents would be proud to call their own. Mark's work at the Welsh Agricultural College had confirmed a natural empathy with the land and livestock, and Paul was about to embark on a six year programme that was to provide him with training for his chosen career that was second to none.

To say we'd lived a simple life would be something of an understatement. Every day had been an adventure, and each setback successfully overcome had helped us all to grow as individuals and as a family. They were hard days, but they were good days. They gave us time to recharge our batteries after a long period of trauma and pressure, and they allowed us to mould ourselves into the ways of Nature again. They were days that we'll always remember with nostalgia and affection.

So it was out of a need to progress, rather than from domestic choice, that the decision was made. The search for our ' Mighty Oak ' had to start all over again.

Once again we made our requirements known to every agent in the area, each one assuming in a wildly optimistic but ill informed ignorance that huge sums of cash were burning equally huge holes in our pockets, and treating us with the exaggerated deference and respect bestowed on all who may be persuaded to part with their wealth. Sale particulars duly arrived by the sackful. A quick look at one or two that seemed on paper to be remotely suitable proved in reality to be a waste of time. We would have to wait, and force ourselves to believe that the right farm would present itself in its own time. Meanwhile there was still day to day work to be done, although now it could be done at a more leisurely pace.

There was one neighbouring farmer that we never did meet, and knew nothing about him until a local property agent had reason to tell us. Apparently he was a tenant who lived on his own and had come to the conclusion that fifty years' hard labour struggling to maintain a lifestyle not far removed from poverty was a big enough price for anyone to pay for the privilege of living in the country. It had been a big wrench leaving his home, but he just couldn't cope any more and had given up the farm and left for parts unknown.

Our informant was related to the owner's brother by marriage, or by some other similarly tentative link, and it soon became apparent that he had every intention of exploiting his family connections to maximum effect. He'd reasoned that, with a bit of luck and low cunning, he could complete a sale of the farm before it was torn from his grasp and offered on the open market. All being well, the resulting commission would be in the bank before any unscrupulous competitor had even got a sniff of it. He'd quickly weighed up the options and decided that we were the obvious buyers, urging us to at least have a closer look. From what little we already knew, it was certainly worth more to us than anyone else but, at risk of upsetting his little plan, we weren't really sure that we wanted it.

It always goes against the grain to agree with anything that agents have to say, especially when you know from experience that the only acceptable motive for opening their mouths, other than for eating, drinking and breathing purposes, is to pursue a course that will line their own pockets with the buying public's money. But we had to admit that the prospect of growing without having to move house was tempting and intriguing. It was reluctantly conceded that he could possibly be right, it would do no harm to walk the land and weigh up the potential.

Judging from the state of the land that bordered us, the tenant must have lost interest a long time before. It was a mess, not too dissimilar to our own land on the day we moved in with our little caravan. His few sheep still grazed the land, and remained well fed and healthy by the simple process of walking through derelict fences and scavenging isolated pockets of half decent grass that peered shyly from beneath an abundance of thriving weeds across the whole farm.

The farm access was on a road running at right angles to our own and a good two miles away, and the land met ours at the extreme end of each holding. Put the two together and the result would be an L shaped farm, two houses and two points of access. The little house and stone cow sheds were plunging headlong into decay, and swift action would be needed if the degenerative process was to be reversed before it was too late. There was no sign of a muck heap and our educated nostrils could detect nothing to suggest that livestock had congregated in the vicinity for at least a full year. Two small paddocks on either side of the house were the only fenced fields, and they hadn't been grazed for many months. Grasses that were already past their best had seeded and wilted to present a depressing picture of total neglect, sweetened only by the overpowering smell of mint that grew in huge healthy clumps around the buildings.

Our initial reaction was that we could fairly simply create a useful holding that could be made viable in every sense - depending of course on the initial cost. The end product could never hope to match the clear vision of our Mighty Oak even if everything worked out well, but it could be a sensible way forward, another progressive step towards our goal. For one of the few times in our latter lives, commercial logic was beginning to emerge victorious. The possibilities had to be looked at more closely.

We learned from the agent that his uncle by marriage, or whatever the owner was, lived at some obscure location in the south of England and seldom, if ever, visited our part of the world. His interests locally were to a great extent looked after by his brother who lived not too far away from us on the outskirts of a nearby village. That seemed a good place to start and an appointment was made for the following day.

The address that was enthusiastically provided by the agent had a ring of grandeur about it, a hint of days gone by when English wealth controlled the more remote parts of the British Islands, much as it did in the colonies overseas. I have a tendency to react badly to the patronising pomposity that has been known to accompany inherited affluence. It sometimes has a restrictive and destructive influence on beneficiaries, and can induce a delusion that effort and growth are only for those of lesser financial standing. The fine line between confidence and arrogance becomes obscured by prejudice of course. I looked forward to the pending visit with some trepidation.

We found ourselves driving through a beautiful rural valley of deciduous woodland and sparkling streams, and eventually to a road bordered by the high stone wall that we were told surrounded the house we were looking for. My negative feelings were reinforced. It was the sort of wall that brings to mind vivid pictures of the lives of the men from a bygone age who built it. They could have been Napoleonic prisoners of war, but more likely they were estate workers or tenant farmers labouring for little more than the food that would provide enough energy to finish the job. Compared with those men, our lives embodied unbelievable comfort and security.

At a bend in the road a pair of high and beautifully ornate wrought iron gates hung securely on massive stone piers to give access to the mansion beyond. We drove regally through the fortifications, our ageing Land Rover announcing its arrival with a cloud of black exhaust smoke and the clatter and thump of a terminally ill engine crying out for retirement. A gardener in dirty overalls emerged from an enormous ivy clad greenhouse carrying a number of potted plants in one hand and a small spade in the other. He smiled and made his way in our direction to introduce himself as the owner of this modest abode, and the man we had come to see. He was as 'down to earth' as we were. I began to relax.

We were guided through the house to a palatial yet cosy sitting room, and invited to take our respective seats in luxurious armchairs that had obviously seen many years' hard work but showed not even the slightest hint of wear. He was a charming man and a most genial host, chatting away about everything and nothing as his wife dispensed tea from a bone china teapot into delicate little cups on a silver tray, as far removed from the gargantuan mugs of steaming brew at home as it's possible to be. The conversation was pleasant and relaxing, though only once did it very briefly touch on the purpose of our visit to provide an assurance that it was in order for us to examine the farm more closely with a view to purchase.

The man was a natural and dedicated horticulturist who lived and breathed for the floral delights residing in his conservatory. We gained the impression that the highlight of any day was to treat invited guests to a half hour shelf by shelf recourse on their origins, beauty and life expectancy, although we found most of them to be little different to the robust spring flowering weeds that decorated the boundary banks of the majority of grazing fields around our home. Sited away from their natural environment, they tended in the main to be shamed into insignificance by bluebells, violets and a multitude of other wild flowers, including the humble ragwort - ignoring for the moment that the latter is a deadly poison and therefore attacked by farmers as soon as it makes its unwelcome presence known. In the circumstances it would not have been prudent or courteous to express such an opinion, so before either I said the wrong thing or our tour was extended to the main greenhouses, we politely took our leave.

Within a week we had confirmed our initial thoughts and tabled an offer for the freehold. Our figure was fair to the point of being generous, so we weren't too surprised to learn that another offer, made a few days earlier, was lower than our own. The agent too dismissed the lower offer, and took it upon himself to send us the usual formal confirmation of sale requesting ten per cent of the proposed sale price and the names of our legal representatives. We in turn notified our solicitors of the verbal agreement to purchase and, in line with the usual cat and mouse games played in such circumstances, assured the agent that the deposit would be paid when the Title Deeds had been submitted for examination and approved. We could do no more than wait for the legal people to justify their exorbitant conveyancing fees by progressing matters at the customary snails pace towards completion.

Everything had happened so quickly and so smoothly, it was difficult to believe that in a couple of months we would more than quadruple the size of our holding without any of the hassle associated with moving livestock, implements and an accumulation of personal possessions, umpteen miles to another farm. It was all a bit too easy and the nagging feeling that something was going to go wrong wouldn't go away. Life seldom hands us our needs on a silver platter and as it turned out this wasn't going to be the exception.

The expected blow came a week later. Wealthy men aren't normally renowned for their generosity or for displaying moral standards beyond the level necessary to maintain a desired reputation for political gain, but the owner turned out to be very much like his brother, a gentleman in the true old fashioned sense of the word. Either that or the agent wasn't as well respected within his wife's family as he liked to believe, and they wanted to give him a sharp reminder that he shouldn't try to be so clever. The news was broken to us in a phone call from the rather subdued agent. The owner felt that the person who had made the first offer should have the opportunity to complete the purchase, irrespective of the sum offered. If the deal broke down, he would be pleased to sell his farm to us for the lower figure that he had decided to initially accept.

As can be imagined it took some time for the reality of this bizarre outcome to sink in, and it's only with the passage of time, and since the misty events of the future have achieved a degree of fruition, that it's possible to look back with satisfaction and in the certain knowledge that it was the best thing that could have happened to us. But at the time we were devastated.

Meanwhile, our growing independence coupled with moving into the house when we did were proving to be both fortuitous and well timed. The end of 1981 saw the worst blizzards to hit Wales in recorded history.

The first one started just over a week before Christmas. The snow was gentle at first, enough to be a nuisance but nothing to worry about. Even so, the ground was soon covered by a two or three inch carpet of fluffy whiteness obliterating every inch of the short winter grass and making any pretence at grazing by the cattle impossible. As usual, they'd been let out for a few hours to stretch their legs, but they'd never seen conditions like that before and an impatient queue of confused beasts soon formed at the field gate, anxiously waiting to be put back into the comparative comfort of the cow shed.

During the next few hours we watched the weather gradually deteriorate, and by late afternoon a driving wind was hurling thick snow across a shapeless white landscape at an almost horizontal angle. From the kitchen window we could see the sheep huddling closer together in the middle of the field, miserably seeking protection and warmth from one another. They were healthy and tough, they'd faced bad weather before during their early years on the mountain, but we got the impression that they knew this was only the first stage of a snowstorm with a difference. There was something about them that shook us out of our complacency and we realised for the first time that the blizzard could be serious.

Lammy and the two horses had come in with the cattle but there was no room to separately pen twenty more sheep as well. Squeezing mixed breeds and species together in confined spaces would almost certainly create more problems than solutions, yet we couldn't just leave them there to fend for themselves even though Welsh Mountain ewes seem able to survive anything.

The three caravans that a short while ago had been our home had since been either filled with tools and equipment or converted into chicken houses, and we'd moved them from their original site to form a square with the end of the cow shed. And to give a degree of shelter from the elements for surplus timber and other bits and pieces I'd built a rough cover over the top, made up with tubular scaffolding and alkathene sheeting. It wasn't ideal sheep housing, but it was a lot better than an exposed field and would have to do.

Away from the protection of the buildings we felt the full fury of the gale, and squinting through half open eyes, stinging from the relentless onslaught of driven snow, we could barely make out the vague outline of our precious little flock which was by then completely submerged in a growing mound of white camouflage. We had this naive idea that they would be pleased to see us, or at the very least acknowledge our heroic journey through the arctic landscape to rescue them. Not a bit of it. Even those who had drawn the short straw and formed the outer ring stubbornly ignored our shouts and prods and wouldn't budge.

The watching dogs turned away in disgust, except for Tina who gave us one of those disdainful looks that is normally reserved for incompetent novices and complete idiots, and without waiting to be told ploughed her tolerant way towards us, head as close to the ground as conditions allowed and ears pricked. Her low growl could just be heard above the screaming wind. The sheep got the message and one by one broke ranks and sulkily plodded towards their temporary accommodation. Once under cover we were forgiven and they came to life again, shaking themselves vigorously before looking around with an obvious appreciation. Our neighbours may have missed the thickly veiled beauty of the resurrected caravans, but there was no doubt that the sheep were glad to have them.

Back at the house, it was a good hour before we realised that there was no electricity, which was a bit of a blow. The boiler was oil fired but of course it was electrically ignited, so we were almost back to square one again with no heating, no hot water and no lights - and we'd only just got used to having them. We consoled ourselves with the thought that it could have been worse. At least we had the solid security of the house and if the power loss went on for too long we could always wire up the generator again. In those conditions it would have been something akin to a miracle if there had been no line damage somewhere in the vicinity.

Throughout the evening we listened to the wind howling down through the valley and felt the temperature slowly dropping. Bed would be the only warm place, and by the light of candles left over from our early days in the caravans we retired for the night. By the early hours of the morning the gale had died away to a simpering whimper and it all seemed to be over. From the comparative comfort of our bedroom we peered through the window into the light of a reflected moon, not daring to guess at what we would see. It was still freezing hard, but for some reason that we didn't immediately understand there was less snow on the fields than when we'd brought the sheep in the day before. There was an enormous sense of relief that our worst fears hadn't been realised.

The next morning we were all up early to check the livestock by torchlight. In spite of tightly shut doors, every animal in the cow shed without exception was covered by a one inch layer of snow that had somehow been driven under the open eaves and through roof vents by a merciless wind. The cattle were confused but nothing more, and after a good feed were relaxed again and back to normal. Dusky had taken it all in his stride and didn't know what all the fuss was about. The thick winter coat of his breed protected him from all the cold and wet that the elements were likely to throw at him, and true to form he seemed to be thoroughly enjoying the discomfort of his companions. The sheep were hungry but otherwise unmoved by the experience and, after filling their bellies with hay, happily elected to file back through the open gate to the fields.

Mark was on morning duty at the Welsh Agricultural College and left for work at about six o'clock. We weren't happy about him driving twenty miles on what could be treacherous roads at that time in the morning but he wouldn't consider not going, and in spite of our concern for his safety we respected his decision.

So with some reservations we followed our normal morning routine, watching the lights of his car move slowly down to the end of the driveway and waiting by the door for him to turn into the road to disappear out of sight. After a long pause he reversed back towards the house. Unseen in the pre dawn darkness, the end of our driveway was marked by a solid wall of snow more than six feet high.

As the new day cast its first glimmer of light across the land, we were able to see for the first time the extent of the blizzard's work. The narrow lane had disappeared, buried to the level of the banks and high hedges on both sides by solid, compacted snow. The only way out was on foot across the fields, and that would have been pointless - and risky to say the least. We were cut off and isolated, except for the telephone which for some unknown reason was still working.

We took stock of the situation. There was a three or four months supply of animal feed, either in the cow shed or securely sheeted outside, so there were no worries in that direction. We had a freezer full of food, plenty of flour for bread, and fresh eggs and milk every day. And of course we still had the machinery to dig ourselves out. We could happily wait for as long as necessary.

Within four days we'd cleared our lane as far as the main road, which had already been rendered usable by Council snow ploughs, and the electricity supply had been connected again. Apart from being a bit cold, the storm hadn't created any insurmountable problems, in fact I think we found the experience more than a little exciting. It's not every day that you find yourself shut off from the rest of the world. At that time of course there was no hint of what was to come in the new year.

Three weeks later, in the early hours of the morning, the second storm hit us. Only this time there was at least twice as much snow and twice as much disruption. It was eleven days, with temperatures well below freezing, before we'd cleared the roads well enough to take the chance of driving into the village. Once again we had plenty of food so it didn't upset our daily routine too much, but others weren't so lucky. Many hundreds of sheep perished in drifts sometimes twenty feet deep, and military helicopters became the only means of feeding the survivors. Milk from the dairy farms was poured into drains. Sick and injured people were airlifted to safety and medical care. Wales came to a stop.

It was once in a lifetime weather. Nature had crept up on her subjects to remind them of her other, more harsh and cruel side. We couldn't help wondering how we would have fared in our little caravans if it had happened a couple of years earlier. I like to think that we wouldn't have been beaten, and that somehow we would have retained our independence and battled with the elements on a day to day basis until it was all over. We'll never know for sure.

Chapter Ten

A Mighty Oak

We visited a lot of farms over the next year or so, looking at farmhouses, cowsheds, hay barns and anything else that was part of the fixtures offered, from the new and modern to the old and near derelict. We accepted generous helpings of tea and refreshments from hospitable farmer's wives, and tramped for miles over fields of every type and standard known to man. But few holdings matched the crucial criteria, and those that did were either too expensive or just didn't feel right. We exhausted the supply of available properties from every agent in the area and combed the "For Sale" sections of all local and national agriculture publications, all to no avail.

Our search centred around the Teifi valley and the village of Llanybydder in particular. It was an area that had held an attraction for more years than I cared to remember and we were frequent visitors to its popular livestock marts, sometimes to buy or to sell but more often to keep abreast of current trends and prices. On each occasion we would call in at the auctioneer's office to ask if he had news of any local farms soon to be put on the market. It was a routine enquiry which always received the same negative response.

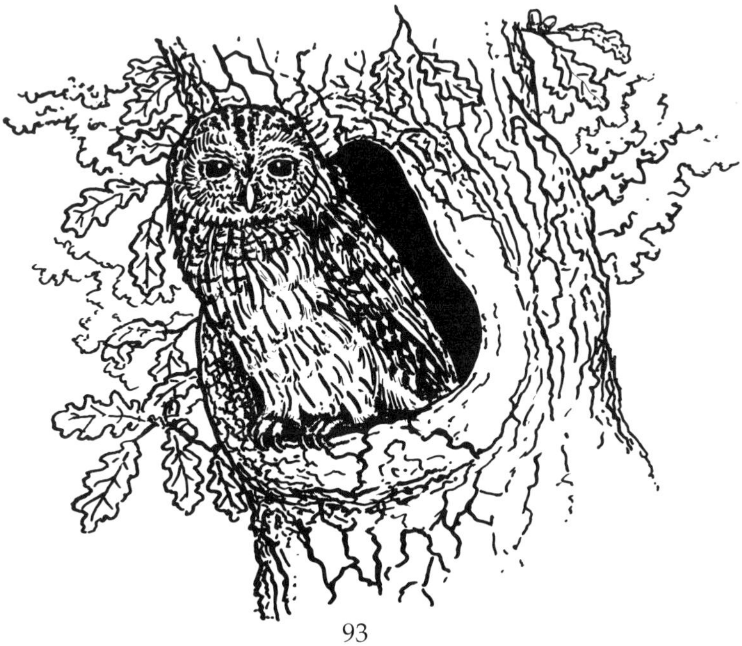

We had to keep reminding ourselves that somewhere out there the farm of our future was just waiting to be found and that we had to be patient and keep looking, but in our weaker moments the black cloud of the earlier fiasco temporarily descended on us and we quietly feared that it didn't exist.

It was in this frame of mind that we once again indulged in the monotonous habit of visiting the mart office to pose the now familiar question. There was nothing special or unusual about the start of the day, but unknown to us it was to be a day that marked a major turning point in all of our lives.

We'd received the usual reply to our standard enquiry and were headed towards the door when the agent suddenly jumped to his feet and called us back. Mumbling incoherently to himself, he disappeared into one of the dusty old filing cabinets that filled the tiny room and emerged a few minutes later waving a grubby sheet of typewritten paper triumphantly in the air, his red face beaming with boyish pride at an unexpected achievement that must be shared. He pointed the paper in the direction of two chairs, which we took to be an invitation to sit down again, and perched his modest frame on the edge of a large and well worn leather chair behind his desk, wriggling a few times to engage the relevant part of his anatomy with corresponding indents in the seat. A bony finger repeatedly jabbed at the precious document as he groped for suitable words of explanation.

It transpired that a local landowner had called in at the mart office more than two years before to discuss his plans to sell one of his farms. He'd originally bought it by capitalising on an amalgamation grant that was available during that optimistic period of twentieth century history when the government of the day had decreed that big was beautiful and infinitely more profitable. Only the land was important and the hundreds of farmhouses across Britain that were subsequently abandoned to the elements were considered to be politically irrelevant in the pursuit of greater material wealth beckoning from the horizon.

So in his retirement years, spent comfortably in a large house in a nearby village, he was facing a need to reverse the whole process and to translate some of his assets into capital. A simple enough arrangement normally but he was an independent and stubborn man and refused to accept the agent's assessment of its worth on the open market. He wouldn't understand that his ranched land and abandoned houses did not combine to realise top values and insisted on an asking price compatible with the better maintained holdings in the area.

The agent had obediently prepared sale particulars but in his infinite wisdom, knowing that the farm would never sell at the figure asked, had committed them to the deepest recesses of his filing cabinet and promptly forgot all about the incident - until now.

His description of the farm sounded encouraging, but we'd been disappointed many times before, sale details are rarely an accurate reflection of the real thing. Even so, we had to admit that it seemed to embody everything we'd spent so long looking for, and in a very weak moment told him so. That was a bad mistake. The result was a generous surge of adrenaline racing through his already overworked veins which had the immediate effect of prompting wild and ambitious thoughts of an unexpected sale. His voice rose half an octave and cash register bells were ringing so loudly in his head that we could almost hear them ourselves. He scanned the paper to remind himself of what he was supposed to be selling, and a barely concealed smile creased his face, suggesting that he was already working out a likely commission.

After a suitable pause to compose himself and to dramatise the moment, he passed the details across his desk, hesitated again as though coming to terms with his own decision, then with an obvious reluctance confessed in a shaking voice that he didn't expect us to pay the full asking price - an announcement that undoubtedly hurt him, and is probably unique in the grey world of estate agents. The effort was unnecessary. It needed only a cursory glance at the bottom line for it to become abundantly clear that he wasn't going to be disappointed. Whilst the farm could well prove to be suitable, the price certainly was not.

We were tempted to congratulate him on his sense of humour and go home, but we had nothing else in the pipeline and there was always a remote chance that after two years on the market with an almost total lack of interest in the property, reason would prevail and a realistic offer would be considered. Perhaps it was worth a try. There was certainly nothing to lose by having a look.

We would have preferred to walk the land on our own, but the owner wasn't the sort of man to give strangers free access to his property, and we understood his reasons. So it was arranged that we meet in the village square.

On the agreed day we set out to keep our appointment to view. It was late winter, and one of those wet and windy days that sends sellers of property into fits of depression and normal buyers tend to stay at home, which is exactly what we would have done if the appointment had not been confirmed.

The twenty mile journey was a nightmare. Rain lashed the sodden landscape, driven to a stinging frenzy by angry winds, and water cascaded down the roadside gullies, sometimes spilling out into swirling lakes across the road. Flood waters from the swollen River Teifi covered acres of rich valley grazing that only yesterday had been home to large flocks of healthy and contented sheep, now huddled together in confused groups on the limited higher ground. Driving was difficult and progress was slow, and as we crept through the deserted village towards the agreed rendezvous point our mood had degenerated to the same low level as the abysmal weather.

There was no sign of the vendor, and we wondered if we were the only fools who had dared to venture out on such a day.

We sat waiting for what seemed like an age, although in reality it was probably only a few minutes, and kept ourselves occupied wiping away moisture condensing on the inside of the windows in a vain attempt to create a dry oasis in an otherwise liquid world. Conversation was limited and the constant drumming of rain on the roof began to be annoying as the time ticked by and our host still hadn't arrived. We found ourselves looking for an excuse to abort the whole thing. No one could expect us to wait for longer than five more minutes in those conditions before turning round and going home again, and that idea grew more attractive by the second.

The unilaterally declared deadline had almost expired when we finally saw our escort's Land Rover appear out of the watery gloom, one arm briefly waving through a barely open window as an indication that we should follow. With a mixture of apprehension and dismay we obediently did as we were asked.

Three minutes later we turned off the main road onto one of those narrow winding country lanes that had been created in the distant past by horse drawn carts. This one had now been tarmaced but the high banks and hedges lining both sides, and the occasional stone cottage built close to the road, were vivid reminders of days gone by. In good summer weather it could have been almost pretty, but it was too congested for our liking and wasn't what we'd been looking for. Once again there was a very strong urge to turn back and put yet another wasted day down to experience, and only the vehicle in front and a necessary courtesy to its driver kept us going. We consoled ourselves with the thought that we needn't stay long.

After half a mile the lane ended abruptly at a rusty farm gate and the convoy came to a halt. We'd arrived at our destination and we weren't impressed. The agent had told us that the house and yard were nearly half a mile from the farm entrance and that the access track was in need of some attention. It was already apparent that his warning was something of an understatement and we prepared ourselves for a long bumpy ride.

As we picked our way carefully and miserably along the rutted surface the rain stopped for a brief moment, the clouds parted and sudden uninterrupted daylight cast a revealing glow over the dripping landscape on all sides. The vision that opened up before our eyes was a never to be forgotten scene only before pictured by a precognitive imagination, and our change in mood was instant, bordering on the euphoric.

Ancient oaks and rocky streams exuded a stability that only time and nature can create, and we could see little that had changed since order was wrestled from the virgin hill countless generations ago. Neglect was everywhere but it was irrelevant. Fences had degenerated into tangled webs of rust clinging tenaciously to sprawling trees, and shallow lines of silt were all that remained of the once effective drainage of the nearby land. But through the chaos there was a kind of order that can only be created by evolutionary process, an artistic talent far beyond the reach of man.

We were completely captivated by the raw beauty and knew instinctively that this was the farm we'd been searching for.

There was a renewed energy born of optimism and confidence, and it took a great deal of self control to temper our feelings with the necessary caution lest we be disappointed again. It had to be remembered that the farm was grossly overpriced and, in the interests of survival, emotions must come second to commercial realism.

Rain was again pouring relentlessly from the sky as the owner pulled into the yard and, jumping from his Land Rover with an athleticism that

probably surprised even him, quickly disappeared through a rickety door in the front of a long stone building which we assumed to be a cow shed. We weren't in any hurry to follow. The moment had to be savoured, and we lingered to cast covetous eyes over our surroundings and to absorb the overpowering atmosphere built up by centuries of human love and endeavour. The farm was obviously very old.

The stone buildings were few, but they had no doubt faithfully served their intended purpose for a succession of residents and their workers. The yard had been enclosed on all sides at some time, and stone gate posts at each end marked a choice of access. The ruined remains of a building, which must have been the original farmhouse, occupied one side of the square behind which was the rick yard accessed on foot through a high gateway topped by a rough stone lintel.

A small detached stable that would have housed three or four work horses during its prime years was flanked by stone steps leading to a loft above, and a large square mounting block, cut from virgin rock, sat solidly by the door, confirming that horses had been routinely ridden in the course of day to day work on the farm. Opposite, two superbly built pig styes and an attached store had been allowed to fall into terminal disrepair, the roof sagging ominously over crumbling walls.

The most recent farmhouse was at right angles to the yard and looked as though it had been built at the turn of the century. It was shabby, but apart from two precariously balanced chimneys it appeared to be in good structural condition, although it obviously hadn't been lived in for a long time.

In the middle of the yard stood a solitary Welsh Black cow.

We were dragged back from the hopes of tomorrow to the realities of today by the sight of the owner peering expectantly from the protection of his cow shed, beckoning us to brave the elements and join him.

We'd never met him before but weren't too surprised to find that his English wasn't very good, although it has to be said that it was better than our Welsh. He was a stocky man in his mid sixties, with the ruddy complexion, calloused hands and rural dress of a typical stock farmer. He was very friendly but surprisingly shy and ill at ease in our company. He told us in halting English that his daughter would show us the farm and act as his interpreter when she came back from checking the sheep. The language barrier prevented much more conversation and we hoped she wouldn't be long.
It seemed an age before we heard the steady thud of a diesel engine and thankfully saw an old cabless tractor chugging in our direction down the track leading from the fields, with a shapeless mound of glistening green oilskins wrestling determinedly with the steering wheel.

The oilskins abandoned their now silent vehicle and moved swiftly towards our refuge, a human face appearing from beneath a discarded hood. The vendor's daughter introduced herself. We guessed she was somewhere in her mid twenties, a cheerful young lady considering her waterlogged condition but with the same underlying caution as her father.

We were reluctant to ask her to walk back across the fields too soon, although she couldn't possibly have got any wetter than she was already, so we were somewhat relieved when she bravely asked if we were ready to look at the rest of the farm, almost certainly not realising that we wanted to see it all, however far we had to tramp and however long it took. The weather didn't matter any more and our surroundings were taking on a whole new meaning.

Field by soggy field we walked, heads bowed against the wind and rain, which is not a bad posture if you want to learn something about the soil, until we reached the highest cultivated land of the farm. Beyond, there was only virgin mountain fronting another small council road. We could see a mass of dense gorse and scrub grass and in the far distance a series of waterfalls spewing out the day's downpour, but apart from providing a second entrance, which could be useful for deliveries needed at that end of the farm, the land was of little value or interest.

It had been a long uncomfortable walk, but well worth it. We were now satisfied that not only was it a beautiful and unspoilt farm but with the addition of a few miles of new fencing and some good management it could also be productive. Most of the land was ploughable, and winter fodder in the form of roots and barley would provide a viable grass break that would also improve the land. Some of the fields had been reseeded over the last few years so we'd have a good start.

By now we were bursting with an enthusiasm that we desperately wanted to hide, and it seemed to be the right time to suggest heading back down to the relative comfort of the yard area to have a closer look at the buildings. Our long suffering guide made a rare appearance from beneath her green hood to enthusiastically agree.

The farmhouse was built to the same basic design that is to be found on hundreds of farms across Wales, and we were encouraged to see that the roof was sound and the day's deluge hadn't yet found a way through to the rooms below. It was bigger than it looked from the outside but badly in need of attention, and although the atmosphere was homely there was an overriding smell of damp neglect and abandonment. A dripping tap over a china sink in the dairy was the single source of water and the only sanitary arrangements were to be found in an old dilapidated privy at the bottom of the kitchen garden. But the house was sturdy and a small fifth bedroom could readily be converted into a bathroom without too much upheaval. There was nothing a lot of hard work wouldn't put right.

Our host was patiently waiting for us in the cow shed, a wry smile creasing his rugged face as the trio of intrepid wanderers squelched slowly across the yard to join him. In our absence the shy and reserved farmer had unexpectedly been transformed into a confident and experienced dealer, employing the same tactics he'd no doubt used for many years in the livestock marts, extolling the virtues of everything he wanted to sell to the point of absurdity. How and why the change took place can only be guessed at, but his new found confidence was a bit disconcerting. He even seemed disappointed that we'd ignored the mountain land, insisting that his sheep always did well up there. We could understand that he wanted us to like his farm but the suggestion that anything could "do well" on a diet of gorse and scrub grass seemed to be taking enthusiasm and salesmanship a bit too far.

The truth was we already liked his farm very much indeed and desperately wanted to buy it. It was also true that if we were to stand any chance of success, his financial optimism had to be dampened at the earliest possible stage, and it was quickly pointed out that a great deal of time and money needed to be spent before the house was liveable and the land farmable, and we would need to calculate the cost of all this work before arriving at any sort of conclusion.

He showed no visible reaction to our comments although his shyness seemed to return. That was a good sign, he was probably prepared to negotiate. But how far each of us was prepared to go remained to be seen.

There was an awkward and embarrassing silence, eventually broken by our host who generously turned the conversation to our existing holding and livestock at Mesen Fach. The polite questions suggested that he was genuinely interested in our past endeavours and we had the feeling that he was pleased that a 'local' family was interested enough in his farm to spend a couple of hours getting wet through walking over it.

For our part it had been a day of emotional extremes. We'd risen from the depths of despair to the pinnacles of optimism, and against all expectations we'd found the farm of our hopes and dreams. We could expect nothing more from a single day and were anxious to leave - in any case it was a safe bet that we all quietly wanted nothing more than to go to our respective homes for a bit of warm, dry comfort.

We thanked the daughter for her help and her father for his time and left to do some hard thinking and calculating.

Chapter Eleven

The End of the Beginning

Our main problems were trying to separate the emotional from the practical, and finding the middle road between optimism and pessimism. We knew that the farm was worth more than our self imposed top financial limit, and that if we wanted it we would probably have to stretch our resources a little. We felt very strongly that we belonged there, but we still had to face the fact that even a revised limit may not be enough and we could possibly end up back at the beginning.

We resolved not to worry about what may or may not happen, and to leave fate to decide the outcome - and our future.

After a suitable period of time we made our first offer. It was low, but not ridiculously so. We didn't think for one moment that it would be accepted, but the vendor had to know that we were seriously wanting to buy, and his reply would tell us if he really wanted to sell.

So for the next few days we waited anxiously for news, pursuing our lives as best we could but finding it increasingly difficult to concentrate on the daily chores that had always given us such simple pleasure in the past.

The tension was broken by a visit from the agent. He was welcome, but we weren't in the mood for unnecessary conversation so the usual polite greetings and preamble were kept to an absolute minimum as we mentally willed him to say what he'd come to say. He got the message and came straight to the point. As expected, our offer had been turned down, but he brought the reply we'd been hoping for. The owner had suggested a compromise figure well below his original asking price, and in doing so had opened the door for further negotiation. We felt an enormous sense of relief, and with it came a renewed conviction that everything would work out well.

From that point negotiations went smoothly and pretty much as would be expected. The only hiccup was prompted by the agent who saw his projected commission getting smaller by the day and tried unsuccessfully to persuade us to move from the figure we had declared to be definitely our highest and final offer.

In the fullness of time everyone concerned was satisfied with the agreed purchase price and we met the owner at the farm to shake hands on the deal in the traditional country way. It was a day we'd looked forward to for well over a year and we couldn't wait to complete all the formalities and to walk the fields of our new home as legal owners.

That should have been the end of the matter, but there followed one of those strange twists of fate that have intrigued the thinking person for countless years, and which on this occasion seemed destined to destroy all of our hopes and dreams. Ultimately everything came together as planned, but our belief that we'd chosen the right course was tested almost to the limit.

Our solicitors were busy following the normal legal procedure of spinning out the whole business to avoid too much embarrassment when the final bill was submitted, and we were working round the clock sorting out a heap of useful offcuts of timber and steel that would have to be taken with us, and another heap of useless bits and pieces that would probably never be used but may come in handy one day so ought to be retained anyway just in case, when we received an unexpected visit from the agent.

He was full of apologies and plainly wished he could have been somewhere else. The message was brief, but its impact was hard enough to make our hearts beat faster and force all thoughts of the planned move into the background. The vendor's solicitors had apparently received an anonymous offer a few thousand higher than our own.

Now we're not exactly naive when it comes to devious business tactics so it has to be admitted that, although it was a bit of a blow, the news was also received with a degree of scepticism. In percentage terms, the new offer wasn't that much higher than our own, and only the vendor, the agent and the solicitors knew our agreed price. So it seemed likely that at least one of them was involved in some way. Was the agent trying to pull a fast one by introducing a fictional buyer to force us to make a higher offer? Had the solicitors been approached by one of their other clients?

Whatever the truth, we weren't prepared to start negotiating all over again and the agent was sent away with a message of our own. He was instructed to inform the vendor and his solicitors that we wouldn't pay a penny more than the agreed price and that if they wished to sell to the new bidder for a higher figure they should get on with it and notify us accordingly.

It was a show of strength and resolve that was a very long way from our true feelings. In truth it was a massive gamble born of a mixture of anger and stubborn intransigence, resulting in long nocturnal hours of self doubt and undiluted depression.

A couple of days later the evening quiet was disturbed by the sound of a car on the long gravel driveway leading to our home, followed by a series of gentle knocks on the front door. We didn't get many visitors during the hours of darkness and any guesses at the identity of this unexpected guest would never have included the gentleman standing by the now open door. The owner of 'our' farm stood before us.

His expression was one of utter humiliation and embarrassment and our feelings towards him softened as he quietly asked if we had time for a chat. We welcomed him as a friend and the kettle was soon boiling to provide the standard remedy for difficult situations, and to give him time to prepare himself for whatever it was that he'd come to say.

Armed with a mug of steaming hot tea he relaxed, and the English words that he found so difficult to grasp came tumbling out in a cocktail of Welsh that left us struggling to understand. Not for the first time, we felt a deep feeling of guilt that we were unable to converse in his native tongue.

We gradually pieced together the reason for his surprise visit and began to understand why he couldn't wait until morning. It became obvious that he was worried and upset, not so much because our attitude to recent events had been hard and unyielding, but because he couldn't bear to think that we believed he was not a man of his word. He wanted to reassure us that the agent had approached us without his agreement, and that he hadn't even considered accepting the new offer. He had shaken our hands on the deal because he wanted us to have his farm and in his eyes such a contract could never be broken.

It briefly occurred to us that we had called his bluff and that he was now trying to recover a sale that was all but lost as a result of his greed, but there was something about his manner that convinced us he was telling the truth. By the time he'd finished his laborious explanation we were believing every word he said and respecting him for having the courage to live by his principles and moral standards. We felt that he was indeed a very honest man, the sort that in this age of selfish materialism is almost extinct.

It was years later that we discovered the source of the bid and the reason for their wish to remain anonymous. The offer was certainly genuine and we were pleased and relieved that our faith in the vendor's honesty had been justified.

He left our home that night a much happier man than the hesitant visitor who had stood on our doorstep a couple of hours earlier. He'd sold his farm for a fair price and was now secure in the knowledge that his reputation was intact. He could retire to his bed with satisfaction and with a clear conscience.

We'd slept like logs as well, but not before our minds had wandered back over the past year or two, pondering the irony and possibly the significance of contrasting and illogical events.

A year or so before, we'd put in the highest bid for a farm that suited our purposes and should logically have been ours, yet through no fault of our own and despite the agent's hard work on our behalf, we'd failed to buy it. And now we'd made the lowest offer for another farm and, in spite of the agents best endeavours to sell to someone else, it would soon be our new home. It didn't make sense and raised many questions for which there were no apparent answers.

But however strange the route we'd followed, it was a heart warming fact that the long search for our Mighty Oak was over, and in the very near future we would begin a new chapter in our lives, one that would prove to be demanding and challenging but which would be faced with the strength of a conviction that we were where we were meant to be.

Chapter Twelve

Moving / Lambing / Shearing / Mountain Land

Moving to our new purchase turned out to be a long and protracted affair. It took a lot longer to complete the sale of Mesen Fach than it did to buy its replacement, and for the first time ever I was separated from Irene and the children, forced by circumstances to revert to a bachelor existence and live on my own at Blaen Wern. We'd made yet another ill advised decision that we subsequently had reason to regret, and agreed to buy the resident sheep as well as the farm. Including our existing flock, we now had something like six hundred demanding 'woollies', all waiting to deliver the fruits of their wombs as soon as the weather plummeted far enough down the meteorological scale to make life as difficult as possible for their lone carer.

We seemed to have inherited a variety of other forms of livestock as well. Big fat spiders had taken up squatters' residence in the house, lurking menacingly in every conceivable angle and orifice and grown to grotesque proportions on nutrients reluctantly provided by a succession of visiting flies who only dropped by to seek shelter from a shower of rain. Those that escaped the strategically woven webs were picked off by families of hungry bats, emerging from the roof they shared with nesting jackdaws through a profusion of access points in tatty ceilings.

I don't mind roughing it but there are limits, and sharing bed space with God knows what sort of creepy crawlies was too far over the top even for me. Eviction from their chosen habitat was obviously going to be a long drawn out operation, and not one that I could easily turn my mind to at a time of mass lamb production. A caravan that had been left at the farm by holidaying friends who wanted an excuse to come back seemed a much better option, although it was like starting all over again, but with a lot more responsibility and without the peaceful nights. At least this time it was a fancy, up market, American caravan, with all the modern conveniences that would normally only be enjoyed in fairly modern homes. But it was still a caravan, and I'd seen enough of those things to last me for some time.

As it turned out I didn't spend much time in it anyway. That first lambing stint developed into a never to be forgotten period of round the clock work with little or no chance to sleep, and snatching a quick bite to eat whenever I could. Time ceased to have any meaning, and I couldn't have told you the day of the week if you'd paid me the price of a shepherd's wages.

The poor fencing, or in most cases the lack of it, didn't help. Despite my best endeavours, it was impossible to implement any sort of system. Each time I managed to separate the sheep into the right groups for lambing some helpful young mother to be would find a hole under the wire and take all of her friends with her. Every flock seems to have at least one fence breaker from time to time, although they're usually sold off as soon as possible. I became convinced that the previous owner of Blaen Wern had made a point of buying every one of them, and my heartfelt thanks and gratitude rang out loud and clear across the fields at regular frustrating intervals throughout the day. In the end I ran out of time and patience and gave it up as a pointless exercise, promising myself that I'd have the last laugh next time round.

And just for the purpose, it was one of those years when a lot of the ewes had twins and needed to be brought in for an extra bit of attention during the crucial first few days. Space in the limited buildings was soon used up and, faced with the possibility of losing lambs outside, I adopted the only solution that I could think of. That was how sheep came to live in the house at Blaen Wern before we did. And very grateful they were too, reclining regally in their straw bale pens in the sitting room, protected from the inclement weather endured by the rest of us, as much food as they could eat and with excellent uninterrupted views across the valley. No sheep ever had a more 'desirable residence', and negative thoughts of envy and resentment were unerringly directed towards the house at frustratingly regular intervals, usually when I was getting soaked to the skin or frozen to the marrow checking on the welfare of their colleagues.

Irene and the children dragged themselves away from the centrally heated comfort of our other home every day to deliver necessary sustenance in the form of ready prepared meals and supplies of fresh food, staying long enough to make sure that I at least ate some of it and then leaving again with any weak lambs that I didn't have time to look after properly for bottle rearing at Mesen Fach.

All good things come to an end, and a couple of months later the poor overworked solicitors had finally got their act together well enough to wrap up the paperwork and we all had a common base again. The intruders in the house were politely asked to remove themselves under pain of undue force, and by the time the inevitable debris had been cleared out the place didn't look too bad. It's surprising what can be achieved with five gallons of disinfectant and an endless supply of elbow grease, and on the day we moved in it was almost suitable for human habitation. The caravan became redundant and was parked up to await its owners' pleasure at some undisclosed time during the summer months.

The two horses approved of their replacement home. Within an hour of arriving at the farm gate they were excitedly exploring new fields and analysing the taste of fresh grass. Helped by Nicky's constant presence, Tammy was now finally and totally convinced that whatever life threw at him, his friends and family would always be there with an abundance of the tender loving care that had been so painfully absent from his early years. Dusky celebrated the occasion by quickly working out how to open the less than perfect field gate and taking his friend for walkies to have a closer look at the rest of the farm. It appeared that some things would never change.

The first major hurdle was behind us but there were plenty more waiting to surprise us at the start of each day, just to make sure we didn't get too complacent. Our living conditions were still less than ideal, but house improvements were gradually pushed nearer and nearer to the bottom of a long list of jobs in need of urgent attention that seemed to grow by the day. Push eventually came to shove and it had to be reluctantly admitted that it would be impossible to do all the work myself, it hurt my pride and offended my independence but if we were ever going to have a decent house to live in again, we would need to employ one or two people on a temporary basis.

It was an uninvited nocturnal guest that finally convinced me. There are periods in the early hours of the night when the brain refuses to switch off, and sleep is only a passing acquaintance. At such frequent times I read and try not to think. Irene, on the other hand, is blessed with an ability to drift into oblivion almost at will, an enviable facility that precluded any possibility of an introduction to our visitor. He came through a hole in the ceiling, completely unafraid but puzzled by the light. He was a pretty little pipistrelle bat. My fascination for these beautiful creatures is not shared by

Irene who has an intense dislike that borders on paranoia of anything with flapping wings that dares to encroach on the domestic scene. So while my nearest and dearest slumbered the slumber of innocence I watched his aerobatic genius at close quarters as he swooped past our heads on circuits of the room and took the occasional breather hanging upside down from the floral curtains. The still, peaceful form next to me would not have appreciated the spectacle. Indeed had she just momentarily stirred, the country peace would have reverberated with uncontrolled shrieks of near panic radiating from somewhere beneath a shapeless mound of shaking bedclothes. It was many years before Irene was told of her close encounter of the bat kind, but the passage of time did not prove to be a good healer.

There was a feeling of relief when the men arrived to start work. For the first day or two anyway. There were two of them, the self styled leader being far more skilful, more experienced and certainly wiser than his quiet partner. Unfortunately he also suffered from an inflated ego and a vivid imagination, both of which were aggravated by a long established verbal diarrhoetic condition to which he seemed totally oblivious but which tended to be extremely painful for everyone else - except his mate who took it all in his stride and just quietly got on with the job of decimating our stock of tea bags.

I've been known to enjoy a nice mug of tea myself, but these two put my not inconsiderable consumption well into the shade. Their limited work time was equally divided between pouring it in at the top end and draining it out at the bottom, suggesting the existence of a straight through tube calculated to cut out all the messy and irrelevant processes in between. Efficiency would have been better served with a funnel and drainage pipe permanently adhered to the relevant aperture and appendage.

The quality of work was acceptable, but only because we were desperate to elevate ourselves back to the domestic standard to which we had only recently become accustomed. With more time and less desperation the oral damage caused by a continual biting of tongues, and the hair loss exacerbated by clumps being forcibly torn from their roots, would have been considerably less severe. Number one builder was blessed with an uncanny ability to bring out the worst in his clients by always believing he was right even when it was patently obvious to everyone that he was wrong, and knowing exactly what they wanted without wasting valuable tea drinking time by asking. Thus there was an occasional difference of opinion and a necessary reminder that we were both occupants of the house and payers of the bills, a privileged, and sometimes painful, position justifying an insistence that the last word was ours. I wouldn't go so far as to call them cowboy builders, though through the occasional thick mist of sleeplessness I was positive that I heard the rhythmic sounds of galloping hooves. It could have been Dusky winding me up again of course.

I was pleased to see them arrive and even happier to see the back of them, but at least we had a degree of comfort and respectability to show for the experience. As the most important room in the house, the kitchen had been singled out for special attention and was at least equal to the demands that would be placed on it at shearing time. On that most strenuous day in the year, food is the second most important input after money. We'd always done our own in the past but this year was different, and we weren't looking forward to it. The thought of shearing six hundred ewes with only one unit filled me with horror. It would take more days to remove that many woolly overcoats than we would care to admit. Our blushes appeared to have been spared when the contractors who had sheared at Blaen Wern for the previous owner turned up to ask if they could do ours. We were only too delighted to book them on the spot.

There were three of them, Neanderthal man and two Goliaths, so I should have smelled trouble straight away. There was little doubt that they would be able to throw our ewes around with the same nonchalance that a modest framed subordinate like me would exhibit in picking up a kitten, and memories of shearing days on Dai and Megan's farm were brought vividly and painfully to mind. Yet I'd still failed to learn from their successful chaos and deluded myself that good organisation would be the key to success. With the self assured conviction of a confirmed simpleton I established a fool proof system solidly based on inarguable logic and naive inexperience.

Adequate manpower was one of the secrets of an effective strategy, and I confidently arranged for two extra helpers to smooth out any unlikely wrinkles in the day's procedures. It would be their job to catch the sheep in pens that had been laboriously set up for the occasion adjacent to the shearing platform in the cow shed, hand them to the shearers as required and then stab the branding iron, liberally coated in sticky dye, on each flank before letting them go. The rest of us would bring in the sheep a hundred or so at a time to keep the pens topped up and, just to be on the safe side, Nicky's friend Annabel and her mother had been co-opted to help roll the fleeces and pack them in wool sacks. It was going to be a hard day's work for everyone, including Irene who would have the responsibility of cooking the gargantuan breakfasts and even bigger midday roasts in her new kitchen that we knew from experience were necessary elements in the employment contract. Everything was well organised, and we were ready to go.

The appointed day promised to be a scorcher, and at eight o'clock it was already quite warm as everyone moved into action like a well oiled machine. At five past eight fifteen naked ewes were trotting down the track, with their bright green brand marks glistening in the sun. It wasn't supposed to happen that fast. At half past eight we'd lost count, the two catchers were already showing signs of dehydration, the pens were half empty, and Neanderthal man chose to drop his bombshell. They wouldn't stop for a cooked breakfast because they had another five hundred to shear on a different farm in the afternoon. I just stared at him, trying to convince myself that it was his primeval way of having a little joke at my expense. An expression chiselled from native stone dashed all hopes, and a slow decline in confidence plunged out of control into a serious nose-dive. The effect of the change of plan didn't need much working out. In about another half hour, at the same time scheduled for the now aborted breakfast, the well oiled machine would seize up. If panic had been in my nature, that would have been a good time to give it free rein.

I'd now learned that the first rule of shearing days dictates that disorder and confusion are indispensable components in the whole operation and that they can only be maintained by doing the first thing that comes into your head, irrespective of the possible consequences. The second rule suggests that the likelihood of a worker's rebellion diminishes in proportion to the degree of fun injected into the delegated chores. So it was that Nicky and Annabel were told to saddle up two horses and take Trixie, the most loyal and hardest working of our dogs, to round up the rest of the sheep and bring them into the yard, while I tried to keep things flowing by assisting with the catching and branding, and Irene used her enforced free time from the kitchen by helping out with rolling the fleeces. All being well we could just about keep up with the shearers, and there was a good chance that our passage through the day could be a lot smoother than with the original plan. It would certainly be a shorter day, one way or the other. It didn't work out too badly until the girls failed to appear on time with the next batch, and the shearers ran out of sheep. They managed to pass the time by checking their automatic counters to calculate how much they'd earned so far, and by drinking a few more gallons of the fruit drinks that had been strategically placed at all points to prevent the helpers' total collapse from dehydration until after the job was done. But we were becoming seriously worried, so it was with some relief that we eventually heard the approaching sounds of two hundred or so vociferous sheep. The delay was thanks to the cattle, who had obviously heard that it was going to be a fun day, and joined in the festivities by breaking through the boundary fence on the mountain land and going walkabout in the general direction of Lampeter. It took a two mile ride in the wrong direction before the offenders were finally located and persuaded to return home.

By some miracle the job was done by lunchtime, and we heaved an enormous sigh of relief. The shearers departed to their next farm, well fuelled with a roast meal of mammoth proportions and considerably richer than when they arrived, and everyone else collapsed in untidy heaps at a number of conveniently available spots around the kitchen.

We found ourselves with an unexpected afternoon to spare, and I opted to comb the farm for any unshorn sheep that may have been missed during the roundup. It gave me the chance to quietly think about the first few months at our Mighty Oak, and what lessons had been learned. We would need to put up some new fencing as soon as possible, so much time was being spent patching up holes that it was interfering with progress on all the other work. And another building was crucial before next year's lambing if losses to the weather and predators were to be kept to a minimum. With that done, we would have some sort of order, and could begin to improve the output of the land, acre by acre. The first year at Blaen Wern had always promised to be difficult but now the worst was behind us.

I found myself at the top end of the farm. I'd never looked at our mountain land before, only briefly from a distance, and it struck me how beautiful it was in spite of, or perhaps because of, the lack of cultivation and attention by previous owners. The land formed the head of a glacial valley, one of many at that height, which eventually join together before sweeping down to our village eight hundred feet below.

The yellow glow of spring flowering gorse which covered much of the south facing slope contrasted perfectly with a kaleidoscope of wild flowers in the shade of the Brechfa forest on the opposite side, and between the two ran the Peithyn Stream, tumbling for a hundred feet in a series of waterfalls that I remembered seeing on the day we first visited Blaen Wern. It came as a shock to realise that although we owned this land, I'd never trodden its soil before, never had time to look and appreciate the beauty of a natural wilderness that had suddenly become very special.

I didn't want to leave, and made my way towards a point where the stream disappeared from view for about fifty yards, and some fascinating slim flat rocks pointed skyward in a double line down to the water's edge. They obviously formed an ancient man made pen, and its purpose could only be to funnel sheep into a deepened section of the stream to wash lice and other parasites from their fleeces. Walking through the pen I came to the dip itself, edged with more rocks and strategically placed at the bottom of a five feet waterfall. Above were more, smaller falls cascading towards me like a scene from a picture postcard.

I sat on a rock to quietly absorb everything that I'd found. No buildings or people were to be seen and I was alone to savour and enjoy an atmosphere of peace and tranquillity that defies mere words. A tired body experienced an elation that can only come from within, prompted and nurtured by the natural magic of an undesecrated environment.

My thoughts turned to children, born and raised in city confusion, who would learn so much about the real world at a place like this. Their blinkered view of life could broaden to embrace a better understanding of priorities and of people. But few would ever have the chance to find out. One day I'll do something about it.

Refreshed and enlightened, I slowly walked back down through the farm to the modest comfort of our new home.

Chapter Thirteen

Quackers / More Dusky / Motorbikes

During the following year we gradually came to know the farm a lot better and hardly a day passed without some new and beautiful corner being discovered in the course of our day to day work.

We explored the oak woodland on both sides of the rock strewn stream that tumbles noisily for half a mile through the farm near to the house, and marvelled at the power of water as bed and banks changed their shape almost daily. We trod fields cocooned by multi-greened trees of every shape and type, grateful for the protection of their unchecked bulk and wallowing in a new and calming dimension to our work. We found springs, gently pumping the purest of water to cool and refresh cattle and sheep and, when toiling under a hot sun, us as well. On the other side of the farm, we walked the length of a smaller stream flowing through the wilderness of a steep sided cwm, alive with the music of countless birds and home to the ubiquitous rabbits.

Wild creatures were everywhere. We watched glistening trout dart from pool to pool and balletic squirrels leaping gracefully in the countless trees. We saw heron, preening themselves on land to the front of the house, and the rare red kite circling overhead. Foxes, badgers, buzzards and kestrels, they all lived their separate lives, accepting our presence in a naturally balanced land of plenty.

Even the discarded mountain land gave up its secret bounty, giving health and bloom to the sheep that grazed there. We didn't know why and, in the absence of any logical alternative, assumed that the virgin land was rich in beneficial herbs and minerals. Silent apologies were offered to the previous owner for doubting him. We grew to love Blaen Wern more with each passing day.

The animals gave the impression that they loved the place as much as we did, or perhaps it was just that the grass tasted that much sweeter. The one possible exception was Quackers who seemed to be a bit lonely in her old age. There was no doubt that she missed the duck pond. It had become her private and jealously guarded domain at Mesen Fach. She had the stream of course, but her ageing legs didn't make much impression on the sometimes strong current, and more often than not she was washed downstream to eventually scramble ashore quacking loudly and furiously at some unseen person who she was convinced was responsible.

There was a danger that one day she would follow the stream bed and get lost, so in our infinite wisdom, and with our usual disregard for commercial logic, we bought a white Aylesbury drake to keep her company and to look after her welfare. The policy of burdening our animals with stupid names continued and the new arrival was blessed with the title of Sir Francis the Drake. It was love at first sight for Francis and he followed Quackers everywhere, standing loyally close by when the old lady sat to rest her weary legs or snuggling next to her when she slept peacefully in one of the sheds. It was a matriarchal marriage, and if Francis wandered too far away from his domineering spouse she quacked loudly and pointedly until he returned to take up his obligatory station at her side. Not to put too fine a point on it, he was hen pecked - or to be more precise, duck pecked.

It wasn't exactly what we had in mind, but Francis seemed to like his subservient role and they enjoyed a trouble free life together, foraging for anything they could find in the damp soil around the springs and earning their keep by cleaning ditches of the snails that play host to fluke, a parasite that attacks sheep livers and, if left untreated, will ultimately kill them. Occasionally we would see them waddling side by side down to the stream to wash and preen in some isolated pool, before idly drying off in the sun at the water's edge. They were a Darby and Joan couple, enjoying one another's company and making the most of every minute in every hour.

The sad day inevitably came when, after a long and happy life, Quackers quietly departed to that great duck pond in the sky, to swim happily and lazily in the sunny bliss of a duck's Utopia. Sir Francis was devastated. His life no longer had any meaning or purpose, and even the grain that he had loved so much failed to tempt his appetite. Without his other half, he lacked the will to go anywhere or to do anything, and he just stood wistfully in the yard as though waiting for her to come back. We found him one morning, about a week later, in the ruined remains of the old pig styes where he and Quackers had spent so much of their time together. His wish had been granted. He'd gone to be with her again.

Dusky meanwhile was busy leading Tammy astray, and creating as much mayhem as possible for as many people as possible, in as short a time as possible.

As expected, our friends who owned the big American caravan came back to spend a few days with us, and drove up from their home in the West

Country in an equally big American car. It was a seven litre monster, totally unsuitable for narrow Welsh roads, built like a tank and with a bonnet longer than the average British saloon. The black paintwork shone like polished glass and the roof was in padded vinyl held in place by moulded strips of stainless steel. It was their pride and joy.

Dusky liked the look of it as well. More than that, he was fascinated by it from the moment they arrived, and chasing him away only increased his determination to examine it more closely. The owners were a bit too protective for their own good, and were foolish enough to shout and wave their arms whenever he even looked like getting too close. It was the sort of angry reaction that he was used to from us but found difficult to tolerate from outsiders. He took it as a personal attack on his farm status and a piracy of his rights. It was a challenge he couldn't resist, and from that point on it became a contest.

Dusky bided his time and behaved himself for a couple of days, lulling his adversaries into a false sense of security. In his own good time, and in the devious way we had all come to know so well, he watched and waited until the unsuspecting owners were happily reclining in their caravan, secure in the naive belief that they'd shown him who was boss and there would be no more problems with troublesome horses. Then he quietly returned to have a closer, uninterrupted look at that intriguing vinyl roof.

It didn't quite work out as planned, and he was spotted through the window. Out they rushed again, shouting, gesticulating wildly and stamping their feet in uncontrolled anger, just as Dusky was enjoying his first full bodied sniff at the vinyl and drifting peacefully away into ecstasy. Understandably, the unwarranted intrusion into his personal and private pleasures took him a little by surprise, and it was more a spontaneous reaction of self defence than malicious intent that prompted him to grab a mouthful as he jumped back.

As I write, those same friends are with us again, and they have the same big American car. These days its roof is decorated with two more stainless steel strips, one to cover a gaping hole and the other to balance it up. They treat him with an exaggerated respect now, voices are barely raised above a whisper and arms are held tightly by their sides. But Dusky isn't interested in their car any more. The vinyl roof has lost its appeal. In any case, he knows that he won and he never bears a grudge.

He was always inquisitive by nature, and had to investigate anything new, or more especially anything new that he knew he wasn't allowed to investigate. It added to the spice of life, and if he could make someone angry and frustrated in the process, so much the better. He couldn't help himself, it was the way he'd been put together.

He'd been dying to have a closer look at Irene's new kitchen for some time, but we'd fooled him by fitting a bull ring type handle to the back door, the sort that has to be lifted and turned all in one motion. A Dusky proof handle. It took him a few weeks but he finally worked it out. Once again he waited until we were all out of the way before satisfying his pent up curiosity and walking in, carefully shutting the door behind him.

He was still there when Irene arrived home, trying to work out how to open the door from the inside. Just to relieve the boredom while his subconscious mind chewed over the problem, he'd changed the furnishings around a little to suit his own peculiar tastes. The potted plants had been pruned back and reduced to pathetic stalks, chairs were tipped over and plates removed from cupboards and stacked in broken heaps across the floor. He'd somehow managed to remove the sealed plastic lid from the sugar store and had gorged himself on its contents, leaving behind a vast sticky mess across the polished slate work surface. To complete the chaos, he'd been unable to find his way through the living room and up the stairs to the bathroom, and had left his calling card in three distinct heaps before tiring of its ordered symmetry and kicking it about in all directions. As I understand the outcome, he was instructed to remove himself forthwith, and dared to return under pain of consequences too horrific to record here. Dusky was in trouble yet again, and very happy.

Whatever his shortcomings may be, and the list is endless, a lack of ambition has never been one of them. One way or another he explored most of the farm, sometimes by mutual consent but more often it was the result of a unilateral decision. But there was one bit of greener grass the other side of the farm gate that he'd never examined, and it looked very interesting and inviting. Pulling back the securing bolt was a piece of cake to Dusky, it was a simple flat plate with a convenient ring on the end, just big enough to accommodate a horse's nose. The first we knew about it was a phone call from a local resident to ask if we'd lost two grey horses. Visitors still sometimes ask why the farm gate is complicated by a safety catch. Only those who already know Nicky's special little friend would really understand.

Dusky is still with us, and shows no sign of going anywhere. He's not the sort to die. He doesn't believe in it. If the decision is left to him, he'll still be around when I'm long gone. I have this recurring vision of Dusky standing astride my grave at Blaen Wern and urinating copiously, making a point in his own inimitable way and trying vainly to make me angry.

In keeping with the traditions of local history and culture, the two horses were sometimes dragged away from their hectic round the clock sequence of eating and sleeping to earn their keep for a change. Usually it was to help bring in the sheep for their periodic pedicures and the slaughter of intestinal worms and other parasites that infest the bodies of farm livestock. They loved it as much as the riders, especially Tammy who needed very little excuse to stretch his legs into a full gallop.

In this way the children became proficient riders, and their endeavours to imitate the skills portrayed by cattle drovers in western films met with some success, albeit with a generous spattering of failures. Whenever possible, their negative experiences remained private and confidential, not to be related to their peers for fear of a sudden deflation of an already delicately balanced ego and a long term image reversal.

Paul took some time to master the art of springing onto a horse's back with the well timed leap that creators of western films had convinced him was the only true measure of good horsemanship. He had been led to firmly believe that stirrups were conveniently and strategically placed only to minimise the energy requirements of the infirm, and to simplify the mounting operation for less competent horsemen and for sissies.

Good timing and the agility of a born athlete are the necessary skills, and if the degrading public spectacle of abject failure is to be avoided, practice in a secluded spot away from prying eyes is essential. There were no doubt many mistimed leaps that ended in flailing arms and crumpled heaps of human fortitude in the grass before I happened by chance to be walking unseen through the field adjacent to Paul's chosen practice spot behind a line of dense trees.

My arrival coincided with yet another attempt to perfect the macho mount, and land solidly and securely in the saddle. I assumed that earlier attempts had consistently failed as a result of too little effort being put into the jump, since this time the enthusiasm and vigour behind his sudden launch into space was spectacular to say the least and did him great credit.

Unfortunately the vigour was a little overdone and he sailed gracefully over the top to land in an embarrassed heap on the wrong side of the horse. Tammy slowly turned to stare disdainfully down at the failed cowboy and, deciding that some human games are beyond the understanding of a mere horse, took two paces forward and tugged hungrily at a particularly tempting patch of grass.

Furtive glances in all directions wrongly convinced Paul that no one had been watching and he calmly mounted in the conventional manner before trotting off back to the house. I wanted to applaud his perseverance and sense of adventure, but kept a respectful silence for many years. He'd suffered enough derision from his brother and sister the last time he was seen to fall off a horse.

Sometimes the horses were left to graze in peace while a more modern form of transport took the strain. The boys both had motorbikes by this time, and they added to the fun of living on a bigger farm. They were useful too, and I must confess to a reliving of my own youthful thrills on two wheels, borrowing one as often as possible to attend to some urgent and mythical job on the other side of the farm. There was a legitimate purpose on occasions as well. A lot of time and leg work was saved checking the stock during the very busy periods. The sheep soon got used to the noisy and unconventional arrival, but the dogs had mixed feelings about it.

Tina was as slothful as ever and heaved a sigh of relief each time a bike was wheeled out of the shed. She was more than happy to sit in the yard and let someone else do the work. Candy had much better things to do than run after motorbikes. She had found her true purpose in life, her raison d'être. She'd noticed that the odd spider was creeping back into the house and took it upon herself to get rid of them. It wasn't an easy job, but somebody had to do it. So she spent most of her time lying in wait by the skirting boards, ready to pounce like a true predator. She usually missed, but she could never bring herself to admit that she'd been beaten by an insect and, with embarrassed glances around the room, sat patiently waiting for the next one.

Pippa loyally followed the bike, struggling to keep pace just in case her assistance was required, but soon found herself two fields behind, with her tongue dragging the ground and in no condition to round up sheep if her life depended on it, although it wasn't in her nature to give in and go home, and every so often I would have to wait for her to catch up.

Trixie loved every minute of it, and seemed to think it was some sort of race between us. Each time I stopped to open and close the gates, she scrambled through some small hole in the fence that she knew about, or jumped over the top to overtake and laugh up at me in that lovely way that dogs do. Sometimes even she got tired, and then she took advantage of one of my periodic stops to jump onto the petrol tank and hitch a ride for the next stage. She was a dog in a million and we still miss her.

Now and again, when all the urgent tasks I could think of that required speedy transport had been attended to, the two boys rode the bikes themselves, sometimes for fun but more often than not for serious work as well. The smaller of the two suffered a serious setback during one day's labour from which it never fully recovered.

Paul was fencing on some of the higher land and decided for reasons that were never explained that it was easier to carry fencing posts and rolls of wire across the handlebars of his bike than it was to use the Land Rover. It all worked without a hitch until a U turn was attempted, and his physics teacher's words of wisdom about fulcrums and resistance were brought sharply and painfully back into his mind. Unable to either stop or reverse the turning action, he motored round and round in ever decreasing circles until a disappearance up his own exhaust pipe became a worrying possibility, at which point he had little choice but to jump off and leave the bike to its own devices. Fence erection immediately tumbled down his list of priorities, and suffered an enforced postponement for a few hours while a disgruntled but much wiser son pushed his wayward machine home and tried to work out why it was refusing to cooperate any more.

Mark was above such foolish actions and roared around the farm on the bigger bike like a true professional, the lessons learned when David came to grief at Mesen Fach just a blurred and distant memory. These things only happen to other people don't they? He'd also forgotten that some of the ditches on the lower land were full of soft, waterlogged silt, and cunningly camouflaged by green moss which Nature had decreed should be only a slightly different shade to the surrounding grass and therefore almost invisible. He was dramatically reminded of his memory lapse during one excursion at maximum speed across the fields below the house, and discovered in the process that motorbikes have a lot to learn from hovercrafts. The bike came to a sudden and unplanned stop as its front wheels ploughed down through the silt, and Mark emulated his brother's developed flying expertise by sailing gracefully over the top in a perfect horizontal racing dive to land at full stretch in the soft peaty soil. On this occasion neither the bike nor the rider came to any great harm, although Mark found it necessary to add a new watch to his Christmas list.

By now a degree of order was slowly emerging from the initial chaos and much of the farm work had become routine, although a lack of stock housing was adding hours to every day and adversely affecting the lambing percentages. Using the house to pen ewes and new born lambs again was a proposal unlikely to be too well received, so we built a new sheep shed to give ourselves another four thousand square feet of cover to protect them from the weather and from predators.

Mark was now working at home full time and a lot of the pressures were lifted from me. There was more time to think, to be mentally absorbed in matters other than farming, and the idea of bringing city children to Blaen Wern came back into my mind.

Chapter Fourteen

Planning the Centre

In those early days plans were vague to say the least. All I had was the thought that children should have a chance to experience the sort of natural beauty that was now a daily part of our lives, and if that was to be made possible at Blaen Wern, some sort of accommodation would have to be built on the mountain land. It didn't cross my mind that it should be anywhere else. I couldn't forget the feelings I'd had a year or so before, the place had a magic about it that was difficult to define. It wasn't just me that felt it, visiting family and friends needed very little excuse to visit that part of the farm. They said it was easy to forget life's problems up there.

Logistically it was the perfect place as well. It had its own road frontage, there was high voltage electricity nearby on our own land and we knew by a permanent wet area on the side of the hill that there was a reliable spring to provide all the water we'd need. In theory it was possible to bring the basic idea to some sort of fruition - provided that the thousand and one potential problems could be resolved, starting with planning permission.

An appointment was made to discuss the whole idea at the Planning Office. Now it has to be said that public sector bureaucrats are not usually my favourite people, especially those whose inflated notions of their own importance prevent them from understanding the difference between power and responsibility, and they seem to be in a majority - or have I just been unlucky? So as I waited for the gentleman to show himself five minutes after the agreed time had passed, thoughts of forcibly expressing my opinions to yet another ill mannered and patronising vehicle of high self esteem were already taking root in my mind.

As it was I was expecting a battle, the mountain land is deep in green belt country and there would normally be no possibility of planning consent for any sort of building for human occupation. My only argument was that I could legally build anything, anywhere on the farm, for agricultural use, without permission from the Local Authority or anyone else, so if I could put up a new building anyway why not bend the rules a little and give city children a chance to enjoy the countryside. It made sense to me anyway.

I was still immersed in negative and aggressive thoughts when I was approached by a very apologetic little man in a scruffy suit who introduced himself as the Planning Officer for our locality. I forced a smile to my face and followed him into his office. His attitude was relaxed and disarming, which did nothing to improve my mood. I'd been on the receiving end of those devious and insincere tactics many times before. But he was so friendly and down to earth that I had no choice but to exercise a restraint that I didn't know was in me and wait for more serious provocation, which I was sure would be provided in due course.

When the unexpected courtesies were out of the way I very briefly outlined the draft proposals, explaining that this was a preliminary visit to discuss any objections his office may have before making formal application. To my utter amazement there were no objections. His support shot me down in flames, and it served me right. The 'rule book' was discarded, common sense was embraced, and the poor chap I had branded as yet another little Hitler, and who just a few minutes ago was to be the target of my prejudice fuelled wrath, endorsed the whole idea with an obvious enthusiasm that left me speechless. There were one or two conditions of course, but they were to protect the environment and I was as anxious as he was that the beauty of the area remained unspoiled.

It was a marvellous beginning that gave us the confidence to push on to the second stage with renewed vigour. We'd planned to register the venture as a charity, and that was when we were brought back to earth with a bump. In retrospect we were probably a bit naive to assume that it would be that simple. Owning the farm was apparently the problem, which is a strange irony considering that it was owning the farm that spawned the idea and owning the farm that would make it possible. In the view of the Charity Commission, we had a 'vested interest'.

I knew what they were getting at and wasn't too happy with the insinuations, although I understood that they had to be careful. Anyway, my feelings didn't matter and, since we had no grounds for argument or objection that were based on facts, there was little point in dwelling on a lost cause. The harsh reality was that if we wanted the benefits of registration we had two choices, either appoint Trustees who were not members of our family or give the land to the proposed charity without conditions. Neither option was acceptable and I refused to discuss it further.

Sometimes in life we make decisions that can never be defended by logical argument, and choosing to ignore this setback and go ahead on our own was one of those times. In truth it was probably based more on personal stubbornness and a refusal to be beaten into submission by non democratic bureaucrats than on any higher altruistic motive. In any event, abortion of the ideas would only have given credence to the Charity Commission's objections, and we couldn't allow that to happen. We carried on with our plans.

And then for some reason we changed course.

I've looked back so many times on the period of conception, and to this day I don't know how we came to expand our ideas and consider bringing disabled children to Blaen Wern. The thought seemed to appear from nowhere. There was no reason, no foundation, no beginning and no event. It was something that suddenly happened.

Whatever its source may have been, it was a crazy idea that logically should never have progressed beyond that initial thought. In common with most people who are lucky enough to have strong, healthy children, we knew nothing about the lives and needs of less fortunate people and, if we're going to be honest, any fleeting thoughts in that direction over the years had been quickly dismissed in favour of more pressing, personal matters.

But now that the seed had been so very firmly sown the least we could do was look into the possibilities. We owed that to ourselves.

So we started at the bottom, examining statistics relating to disability in general then phoning and writing to all the major disabled children's charities in the U.K., and subsequently examining the literature of organisations catering for their special needs. We were astounded by the number of children with physical and mental health problems and, even at this early stage, it was apparent that suitable facilities for them were few and far between.

Our eyes had been opened to a different area of life and we realised that we were already getting ourselves mentally involved in a sphere that we still knew very little about. But it was a phone call to a charity head office in London that finally convinced me that we had unwittingly uncovered a very real need. I was asking for information regarding some point or other and was put through to a lady who arranged holidays for disabled children. Her voice was middle aged and her manner a little abrupt, prompting a spontaneous and unfair mental picture of the office dragon.

She said very little to begin with, and I was quietly writing off our one sided conversation as a pointless exercise when my questions unexpectedly received more than her usual curt and monosyllabic reply. For the first time, I noticed that her speech was laboured and hesitant, and it became obvious, even to an inexperienced ear, that she was disabled herself.

It was as though displaying her disability had somehow bridged an imagined gulf between us because she was suddenly relaxed, speaking at length and with an air of complete authority on a subject that she understood only too well. My questions were answered one by one and as our conversation expanded to include the very rough proposals for accommodation at Blaen Wern she grew more and more excited by the possibilities, and more and more emotional.

I knew then that we must at all costs finish what we had only just started, and from that point on research became more serious and committed.

A list was made of all the things that we were told the children were unable to do or to experience, and it was noticeable that their access to the countryside was very severely limited. Phone calls were replaced by personal visits and many miles were travelled to talk to people already involved in some way with the children, although we deliberately avoided contact with anyone providing recreation in a practical way. We wanted no pre conceived ideas.

Strangely, our biggest strength was a complete lack of experience. We were able to look at the overall situation in a fresh and objective way, and throughout the following months it helped us to gradually develop a better understanding of the whole sphere. There were plenty of blind alleys and frustrations but that was all part of the learning process, and even the wasted journeys contributed something to our store of knowledge. In a negative sort of way they were instrumental in arriving at the final basic philosophy, and pointing at the road to the future.

One such journey was to influence the format of a charitable trust formed many years later. It was a five hundred mile round trip to meet a director of one of the largest charities. I must confess that I didn't like the man in the first place, far too arrogant for my liking, and when he told me his home address any remaining respect disappeared. It was an affluent part of an affluent town that I knew well. I later checked on his salary, it was enormous, paid by people with collecting tins on street corners.

Another, shorter, drive to visit a man who had built self contained units for disabled people, and who was himself confined to a wheelchair, proved to be more constructive, although I felt that the biblical quotations were unnecessary. That should be a private matter and people can make up their own minds which spiritual direction they wish to take.

Slowly the final plan emerged from the mass of information and experiences collated over a two year period. We'd tried to fill as many voids as possible and the end result reflected the opinions of disabled people themselves and our assessment of the children's needs, cast in a mould to conform to our relaxed and informal way of life.

If there was one single thing learned from the years of research it was that, although the material needs of people with health problems are as diverse as the British weather, dignity and freedom are always paramount. The Centre was designed to that end, with large bedrooms and en suite bathrooms, open plan lounge and dining room and a heated indoor swimming pool. We'd also included a licensed bar for good measure, and bought a coach to offset the travelling difficulties of the larger groups.

What had begun as a plan to provide basic accommodation for city children had expanded to embrace the needs and wishes of disabled people of all age groups and bore no resemblance to the original sketched drawings. Unfortunately the cost had followed suit, spiralling alarmingly with each addition, until the inclusion of the swimming pool signalled the end of all possible financial expenditure.

So this was it, the end of the preparatory road. In late 1987, four years from the first tentative idea, we were ready to start work. We'd either got it right or we hadn't, and we were about to spend a small fortune to find out.

Chapter Fifteen

Building the Centre / Trevor / More Horses

We couldn't have chosen a worse time to start work on the Centre, it was one of the wettest winters on record and thousands of tons of glacial gravel and rock needed to be dug out of the side of the hill to create a level plateau that would accommodate the building itself.

Our ageing digger was well past that sort of hard labour, it was really a job for heavy earth moving plant - which would be expensive. We'd carefully budgeted every part of the whole programme and the cost of hiring machinery with operators was about the same as the cost of buying another digger and doing the job ourselves. So we crossed our fingers and spent an absurd amount of money on a replacement machine, hoping that the theory would be confirmed in practice. As it turned out it was one of the best decisions we have ever made. After countless hours of grindingly hard work during the whole of the building programme, it had more than paid for itself. We still have that same excavator and it's still earning its keep on the farm.

It was during the early part of the Centre's construction that Trevor became involved. The English manufacturers of the necessary steel frames had run into financial difficulties and our order was long overdue and would in all probability never be met. I wanted steel stanchion housing fabricated so that work could continue while our frame requirements were manufactured elsewhere. Trevor was an old friend from our early days in Wales, and his knowledge and experience in steel fabrication made him an obvious target for advice.

He was a man who cared even less about convention and fashion than myself, heavily bearded and dressed for work at all times he was as strong as an ox and projected the image of someone as far removed from the real Trevor as it's possible to be. The reality was a gentle man with enviable artistic and creative talents that were given the freedom of expression through model making and drawing, and on occasions, when time and motivation converged, through painting in oils. His work potential was severely limited by a congenital heart defect that at times of physical exertion left him short of breath and energy. But I never heard him complain, and he would do anything in his power to help someone in genuine need. His heart may have been faulty, but it was made of pure gold.

Trevor came to our rescue, and not only made the planned steel housing but stayed on to help in every way that he could until after the Centre had been built and opened. Without his help the work would almost certainly not have been completed within budget and timescale. We owe him a lot, and no words will ever fully redress the debt of gratitude felt by everyone in my family and beyond.

As work progressed and the Centre began to take shape we found ourselves thinking more about the other ingredients in the overall plan, especially the animals. We already had the farm livestock of course, and they would all play a part, especially in the spring and early summer months when there would always be a few hand reared lambs that the children could feed from a baby's bottle. The Welsh Black cattle were a bit big and frightening so we'd have to find a way of getting the children close to them in safety. The four collies were friendly souls ready to welcome anyone with an enthusiastic licking as soon as they had carried out their duty and barked furiously to let us know that we had visitors, and to give the absurd impression that they could be vicious just in case the intruders harboured impure thoughts. And we had Dusky and Tammy who were both quiet enough to take children for rides.

But we needed more horses, one or two for riding, and something small and quiet that the children could relate to and learn to understand. We heard about the Shetland from friends of ours. She was a little black mare and belonged to a young couple who lived about twenty miles away. They had hit a cash crisis and couldn't afford to keep her, much to the disappointment of their young children.

It was worth investing a couple of hours to go and have a look, and we're glad that we did. Watching her reaction as the two youngsters crawled all over and under her left us in little doubt that she really was a bomb proof pony, and we bought her on the spot.

We discovered in the process that giving animals ridiculous names wasn't a peculiarity restricted to my family. The poor little mare had been saddled with the name Snugglebum, which the vendors hastened to assure us was their children's choice. Be that as it may, I couldn't see myself standing in the yard calling Snugglebum at the top of my voice. Neighbouring farmers would shake their heads knowingly and sadly, convinced in their own minds that I'd finally flipped. At a push I was prepared to call her Snuggles, but even that was likely to prompt a few giggles and silly questions. My suggestion that we should select a new and more sensible name was received with collective family horror, so the poor mare was stuck with it, and so was I.

It was against our better judgement that we acquired another horse at about the same time. Shandy was Methuselah's father, and what exactly he was doing at Blaen Wern I really don't know. He was bought by a friend who knew we were looking for good reliable ponies, so thought he was doing us a favour by bidding for him at a sale and booking his cost down to us.

The theory, apparently, was that he would make a good trekking horse. The sad truth was that he had possibly already been a good trekker at some distant stage in his long history, but he was by then, and there's no point in hiding from the truth, a broken down old nag.

But we were stuck with him and, needless to say, that unfortunate character trait that had already cost us so much time and money in the past surfaced yet again and we got fond of the old fool.

On the plus side, we had no doubts that he could be trusted with children and wouldn't suddenly take off across the fields. To be brutally honest, he couldn't have galloped across the fields if his life depended on it. We did manage to raise his pace to somewhere between a clumsy trot and a lumbering canter once or twice, but it took a lot of hard work from the rider and would have been a great deal easier and faster if we'd reversed the more usual process and carried him on our backs.

We tried boosting his energy levels with oats and sugar beet, but only succeeded in getting him hooked on it. At feed time he could barely wait to get his nose in the bucket, and skipped around the pen like a yearling colt until the blessed nectar was served up in the usual generous proportions and his withdrawal symptoms were satisfactorily eased.

Sugar beet is an excellent animal nutrition, but one of its more doubtful properties is the ability to swell up to two or three times its dried size when introduced to liquid. That's not a problem for cows and sheep, they just regurgitate it as cud and digest the augmented volume in their own good time. It's a different matter with horses, and a thorough soaking in water for twenty four hours is essential before it can be safely offered. Feed it too soon and a stomach ache is the least of the likely problems. Leave it soaking for too long and the sugar turns to alcohol.

Sometimes too much was prepared, and then the unwanted surplus was moved to the back of the shed for disposal at some later date on the muck heap. Fermenting sugar beet has a smell like no other, and the wafting odour from a discarded bucketful of tantalising sweetness must have driven Shandy crazy with longing one night. Judging from the churned up bedding around the edges of his pen he'd made more than one attempt to get to it before the result of four or five scoops of beet and oats from his evening meal became topped up with a good supply of adrenaline, and he found himself with the energy he needed to break out and satisfy his craving.

Stopping when he'd had enough was never one of Shandy's better qualities, and we found him the next morning tottering around the shed in a drunken stupor and not knowing if he was still at Blaen Wern or winging his merry way to planet Mars.

Those who have themselves been tempted to over indulge in the devil's brew will understand that there is usually a price to pay, and Shandy's fall into temptation proved that he was no exception. Throughout the rest of that day he stood silently staring at the ground, eyes glazed, nose running, and with the look of a creature struggling to come to terms with the king of all headaches.

He survived the experience - just. But he never broke out of his pen again. He wasn't the cleverest of horses, but he was wise enough to learn that the cost of greed far outweighed the short term pleasures. He decided to behave himself in the future, and to make the most of his legitimate rations. And we in turn decided to accept him as he was, to make the most of his slow, but sober, walk, and to keep reminding ourselves that he would be good with children.

A few weeks later fate intervened on our behalf and led three lovely horses, perfect for our needs and with good sensible names, to our door. They belonged to a lady who worked in one of the village shops and who had heard about our plans on the country grapevine. Trina had owned them since they were foals, but a lack of time and available grass was making it very difficult for her to keep them, although she couldn't bring herself to sell them to the highest bidder not knowing where they would end up and what sort of a life they would have. Her suggestion took us a bit by surprise. She offered them to us as a gift, but she hoped she could call round now and again just to say hello. And so, completely out of the blue, we became the very proud owners of Gemma, Bracken and Lucky.

Gemma was a magnificent registered Palomino Arab mare, a little over fifteen hands high and with a temperament to match her beauty. Tammy couldn't believe his luck. It was many years since he had enjoyed the company of a female of his species and he immediately fell head over heels in love, slavishly following her everywhere. His long and close friendship with Dusky became history overnight.

Nine years later and nothing has changed. He can't bear her to be out of his sight, and when Gemma is being ridden Tammy either follows closely behind or races backwards and forwards along the fence line whinnying pitifully until she returns. In winter they have adjacent stables with a mesh window between them so that he can see her at all times. He's still besotted.

Bracken was a Connemara, a tough Irish horse with tight conformation and a thick curly coat, a breed that has served its people over hundreds of years with the same loyalty as the Cob in Wales. He was in his twenties and totally reliable, nothing would spook him because he'd seen it all before. Lucky was his little friend. They'd been together since they were colts and where one of them was the other wouldn't be far away.

Lucky was another Shetland, a bit bigger than Snuggles and with his own private claim to local fame. He was trained to harness and for many years had delighted owner and spectators alike as he trotted around the show ring at local fairs pulling his own special little purpose made trap. He loved every minute of it, an extrovert showman if ever there was one.

And now they were ours, and their new function in life was to make a lot of children happy, and to help them to know and understand horses. We were confident that all of them would do that job well.

Chapter Sixteen

Sam the Shetland / A Dummy Run / Swiftly the Welsh Cob

Horses are a part of local culture, and if our modest little village has any claim to fame, it's because of the horse sales held in the nineteenth century market place on the last Thursday of every month. A hundred years ago it was the indigenous Welsh breeds that attracted the buyers, but these days almost every size and breed of horse, from the roughest workhorse to the grandest thoroughbred, is paraded for sale to visitors all over the world.

It has never been our favourite place and we seldom go, even though it's only a couple of miles down the road. There's always too strong a temptation to turn Blaen Wern into a sanctuary for the horses that we know are likely to be exported to a very uncertain future in Belgium or France. Quite simply we can't afford to buy them all, so we take the cowards way out and stay away.

On one rare occasion Nicky and I decided to have a quick look to pass half an hour that was unexpectedly spare. We'd stopped in the village on the way back from a catering sale, where we'd successfully acquired necessary kitchen equipment for the Centre, so that Irene could attend to the routine shopping requirements of a hungry family.

There were some superb animals on show that day, and we spent a happy ten minutes or so wandering from pen to pen admiring the fine quality of the aristocratic occupants, until I realised that Nicky was no longer with me. I found her standing in front of a little Shetland mare and her foal. They were a mess, fed well enough but they'd never seen the inside of a stable and the elements had welded their coats into a thick hard crust. Small bare patches suggested the presence of lice. They were to be sold as two lots, the mare would probably go to a home at least as good as her last one, but barring a miracle her offspring was destined for the continent.

We tried very hard to forget them but before too long, being a bit soft in the head and with a tendency on such occasions to forget that I was even softer in the back pocket, we found ourselves in front of the same pen again. The little colt's nose was now within two inches of the concrete floor and his body sagged under the weight of insurmountable problems and utter despair. I gently lifted his head but there was no reaction, and dull eyes just stared blankly into space. He was desperately unhappy.

Irene had joined us to complete a formidable combination of two emotional females and a male who was too easily swayed by sad eyes for his own good. I knew then that the pony's future was assured. As the auctioneer's hammer came down on our bid I vowed never to visit the horse sales again, a pledge that was reaffirmed again and again as I made my way to the mart office to pay the bill. I was annoyed with myself. We shouldn't have bought him, we shouldn't have allowed ourselves to get emotionally involved and we probably shouldn't have even gone to the sale in the first place because we might have guessed that something like this would happen.

With negative thoughts still racing round and round inside my head, I made my way back to the pony. Nicky was grinning from ear to ear as she stood holding him on the end of a borrowed lead rein. He was standing patiently next to her, his cheeky little head held high with ears pricked, and surrounded by excited children running from one side to the other, stroking his nose and patting the thick matted coat. He was totally unperturbed and loving every minute of it. I had a very strong suspicion that he'd weighed us up as soft touches and likely benefactors, and that we'd been well and truly conned.

We put him in a pen with Shandy as some sort of compensation for being dragged away from his mother, and knowing that he wouldn't be bullied as newcomers to a herd often are. He settled down straight away and we left them that night happily eating hay and getting to know one another. The following morning the pecking order had been well and truly established and any minor doubts we may have had had disappeared. Sam was stretched out full length in the middle of the pen, in a deep peaceful sleep and watched over by a wide awake Shandy, standing motionless in the limited space that remained in one of the corners.

Today, Sam is probably the most loved and the most photographed animal at Blaen Wern. Like Dusky, he's amassed a vast repertoire of antics calculated to gain maximum attention, and he usually succeeds. He was a fast learner, and would have been a circus owner's dream horse. To prove the point, we taught him to shake hands, rewarding him with a polo each time he did it properly - which was probably a mistake. Sam decided to reverse the process, and whenever he fancied a polo, which was often, he followed us around like a pet dog, raising his right leg in greeting until he got what he wanted.

He's still full of fun and always happy, and we've never seen him get upset with anything or anyone. He has a wonderful ability to make us laugh, although we're never sure why or how. He's one of those precious rarities. He's a character.

Meanwhile work on the Centre had progressed more or less to schedule and to budget, and by September 1988 it was nearing completion, although putting the finishing touches to everything was likely to go on right through the winter. It was fitting really that the first time the Centre was used was for a family occasion. It was my sister's idea, a surprise get together of friends and family from all over the country to mark our mother's eightieth birthday.

It was a good weekend, and in many ways it was also a dummy run for our first big group due to arrive in the spring. That was to be the ultimate test for us, and I have to admit to some last minute doubts. We'd stuck our necks out so far that the Sword of Damocles couldn't miss. A new and straightforward business venture is a big enough risk, but walking a tightrope between commercialism and altruism, with no charity status to fall back on, was madness bordering on financial suicide.

It promised to be a long winter of fluctuating feelings, from periods of buoyant optimism, and confidence in my own irreversible decisions, to dark moods of self doubt and possibly regret. It was perhaps fortunate in a way that another member of the animal kingdom occupied a lot of my thoughts for most of those months, although I was not to know at the time what a profound and lasting effect he would have on my life.

We needed one more horse to complete the range, a strong animal that was trained to harness and would pull a cart, but our efforts to locate one had so far proved fruitless. Our finances were very low, and perhaps we were expecting too much to find the right animal that we could also afford. The best we could come up with was the offer of one on loan, free of charge in exchange for the grazing. Memories of the Domino fiasco immediately came flooding back, but we agreed just the same - on the condition that it would only be until we found what we were looking for at the right price.

He was a Welsh Cob, arguably the noblest of Welsh breeds. They're fast, strong and sure footed, and in days gone by they were used by the military to carry men and heavy guns into battle. In more peaceful times they worked between the shafts on farms or as riding horses over the rough mountain terrain of Wales.

He was a bit lean, and his feathered feet and slightly heavier frame suggested that he carried Fell blood, but his shiny black coat glistened from the regular and vigorous brushing that's usually consistent with good care and attention. His broad back was complemented by a short neck and strong hind quarters, and his tail was shaved and bound with brightly coloured ribbons matching a well combed and equally colourful mane. His powerful legs quivered from a mixture of anticipation and impatience. He was obviously nervous, but that was understandable.

He needed to calm down and come to terms with his new surroundings, so we wasted no time in setting him free in the nearest field. And there he stayed for the next few days, eating the ample grass, sleeping, and most importantly, relaxing.

Left alone he was happy enough, but as the days and weeks passed we came to understand him better and realised that there was more to him than was immediately obvious. He was never at ease with people, especially men, and worse, he was easily spooked, which was a bit worrying and made us wonder if the whole thing was another pointless exercise. Approaching him from any direction except directly in front would inevitably prompt a speedy retreat to some distant point from which he could safely view the intruder.

He didn't seem to mind a head collar, but the sight of a bridle hanging loosely from the hand sent him galloping away across the fields, tossing his head with an energetic show of resentment and anger.

Catching him for riding was a delicate and devious exercise, walking slowly towards him from the front with a polo in one hand and a rolled up lead rein in the other, and a bridle carefully concealed inside a jacket or up a jumper. Once the lead rein was looped around his neck the battle was won and he became as quiet as a lamb, accepting the bridle without further fuss. This ridiculous charade was ritually carried out each time we needed him.

We were puzzled. He loved to run and enjoyed the attention of being ridden, so why did the sight of a bridle upset him? It remained a mystery until the day he caught sight of a visitor carrying a stick. He was suddenly both furious and terrified, his white eyes bulging as he tossed his head and strained frantically against the restricting lead rein, dancing and prancing in endless circles to the rhythmic sound of his snorting anger.

In that brief moment it all became crystal clear. He was obviously terrified of anything remotely resembling a weapon, and through a horse's eyes a loosely held bridle with the reins hanging close to the ground could easily be mistaken for a whip. At some time in his past he'd been beaten, not once but many times, and on each occasion it had been by a man.

There was now little doubt that his early years had been hard beyond belief, and that a hard and uncaring discipline had moulded him into a neurotic introvert who was likely to erupt and react violently to any reminder of his deep rooted fear of men and a repeat of past brutalities.

As with all animals, such early mental damage is irreversible and sometimes can be insurmountable. It would be far too risky to use him at the Centre, the possible consequences were too horrific to even consider. The prudent course would be to send him back to his owner and put the whole episode down to yet another equine experience. Yet underneath it all, I felt sure there was the faintest hint of a gentle and patient creature waiting for the chance to show himself. A creature crying out for someone to trust and to serve, someone who in turn would care about him. He wasn't our horse but it was a challenge I couldn't resist.

The adopted tools of the task were modest. Generous helpings of patience, perseverance and that much loved bribe and reward used by most horse owners from time to time over the years, polo mints.

Swiftly loved polos, and after a few thrice daily excursions to his field, talking quietly to him as he crunched his way through half a packet at a time, he began to look for my arrival with an obvious enthusiasm, and before too long he was plodding his way towards me as soon as he heard the sound of the gate opening. It was a good start and made the job of catching him for riding a lot easier, although we still had to resort to the same underhand methods as before.

As winter drew nearer, all of the horses were taken down to the lower part of the farm ready for housing. Dusky and Tammy were well used to the seasonal routines, and each afternoon, the time depended on the weather, they made their way to the yard gate to wait for each in turn to be led to his or her pen where a good supply of hay and cereal was waiting.

Swiftly quickly fitted in with the others and wasn't backward in pushing his way to the front of the queue. He had a good appetite, and the less charitable among us could be forgiven for suggesting that he was greedy, although he would always stop munching to take a polo - just to aid his digestion, you understand. He knew that I always kept them in the same place, and would very gently nudge the relevant pocket to tell me that one or two more would be appreciated.

It was my habit of lingering with the horses and cattle after they'd been fed in the evening that finally sealed the trust between us. I was used to good natured ribbing about talking to the animals, but this time there was a real purpose, and Swiftly responded to my extended presence as I hoped he would. He welcomed the company and it didn't matter to me that it was probably only because of the extra rations I gave him, it was only important that he knew he'd at last found someone he could rely on. The fact that that person was soft enough to be swayed by a trusting look and some gentle nuzzling was, to Swiftly, an added bonus.

By the end of the winter we had become good friends. I knew all of his little idiosyncrasies and no doubt he knew mine. He was a different horse, quiet, trusting and affectionate, and probably happier than he had ever been. I should have been as happy as he was, but I knew that the bond between us could only be temporary. We'd spent so much time together that I felt he belonged with me, but that wasn't possible. He wasn't my horse and the day must inevitably come when he would be taken away.

That day came early in the spring. It was the first time we'd seen Swiftly's owner since the day he'd brought him to Blaen Wern. Now he wanted him back to sell him. He had no way of knowing the happenings of the preceding months, or of the changes in his horse's temperament, and I didn't tell him. There was no point. It wouldn't have made any difference.

However much I juggled with cash flows and projections, and I did a lot of juggling, there was no way we could afford the asking price, or even the figure I thought could be negotiated. Swiftly's future was out of my hands, and I could only hope that his new owner would care for him as well as we had. I mentally prepared myself for his departure.

I was perhaps too involved in my own disappointment to notice that other people cared as well, and that my sister had been quietly watching as my feelings for Swiftly deepened and as a nervous wreck of a horse grew into a gentle and loyal friend. I was blind to her private determination to make sure that no one would ever take him away. I'd considered the problem to be mine, and mine alone.

Yet suddenly, unexpectedly, and with no opportunity for argument, gloom and despondency were banished by someone who had nothing to gain except the pleasure of making others happy. Swiftly was bought and paid for. He was going to stay with me, where he belonged. I shall always be grateful to Sybil, not only for a gift that would be difficult to surpass, but for caring enough about us both to use her hard earned capital to make sure we'll always stay together. It was typical of her caring and unassuming nature about which I could write a great deal more.

Swiftly has calmed down even more since those days, although he still has his moments. He's a very strong horse and not easy to stop when he gets into his stride, but he's also sensitive and responsive, and loves the occasions when he's used to round up sheep or cattle, using his low centre of gravity to maximum effect, twisting and turning to head off the strays.

Above all, he's very happy in the knowledge that he will be at Blaen Wern for the rest of his life, and so am I. He still plods across the field to greet me as soon as I appear, and he still nuzzles me gently as I stop to talk and to stroke his shiny black coat - he's never forgotten that there's always a good supply of polo mints in my left hand coat pocket.

Chapter Seventeen

Early Lessons / The First Children

And so winter passed into the spring that would answer so many crucial questions, and would in its own way pass judgement on the wisdom of the decision to extend ourselves beyond the cosy realms of a farm kitchen and into the vastly different lives of people we'd never met, people we hoped would love Blaen Wern and the West Wales countryside almost as much as we did.

It was supposed to be a gentle transition, a sort of apprenticeship to help us find our feet and to iron out any wrinkles. For that reason our first guests were families and couples so that looking after their needs was likely to be little different to welcoming friends into our own home, even though we appeared to have very little in common. They were in the main city type people who had lived their entire lives under the shadow of some sort of debilitating illness, and we on the other hand were peasants only recently aware that there were such illnesses.

It transpired that we had one very important thing in common. They wanted us to succeed nearly as much as we did, which as it turned out was just as well.. We were at the very beginning of an enormous learning curve and the help we received from those early visitors was to stand us in good stead for a future that at that time was by no means certain. It was a relief to learn that we had the general design and the bathroom aids about right, but we soon discovered that we had a lot to learn about the everyday practicalities, and our inexperience let us down with a disconcerting bump on more than one occasion. It was sometimes embarrassing although our shortcomings seemed to add to our guest's determination to help and to the friendly informality.

We'd spent a lot of money on the beds, acting on the theory that if people slept well and ate well everything else was likely to fall into place. One small but important point that we overlooked was the recommended removal of castors from all furniture. The oversight was brought sharply to our attention when a newly arrived guest sat on the edge of his bed before unpacking, and was surprised to find himself transported at great speed from one side of the room to the other with no visible means of arresting his progress. He assured us that he found the experience quite exhilarating and likened the sensation to being strapped to a skate board - although how he would know what that felt like we couldn't imagine, and didn't like to ask.

The bar was a well used and popular inclusion in our plans but it was foreign territory to a family of bumpkins who much preferred a pot of steaming hot tea to a pint of bitter or glass of scotch. On very rare occasions we had enjoyed a drop or two of something stronger at home but had seldom found the time, or indeed had the inclination, to visit public houses. The range and content of drinks that we had been advised should be speculatively purchased as stock, and the volume of each one that is normally purchased, were relatively unknown factors.

Irene has an enviable capacity to get on well with everyone, so it naturally followed that she would be co-opted into the role of part time barmaid. She coped well enough with the normal beers and spirits, a pint is a pint and optics measure everything else for you anyway, but if someone wanted a strange complicated mixture of two or three different beverages they either had to do it themselves or give her instructions.

There was one memorable lady who was fairly well hooked on brandy. She was in her late middle age, always smartly dressed and with the most beautiful golden brown hair. When at home she apparently indulged in two or three glasses of her favourite tipple before going to bed, on the pretext that it was the only medicine that guaranteed her a good night's sleep. It was a reasoning that fooled no one, especially the professional carer and friend who accompanied her, part of whose job it was to make sure she didn't overdo it.

The end of dinner on her first night was the signal to remove herself to the bar as swiftly as her less than perfect legs would carry her in order to partake of her routine medicine. She sipped it slowly and deliberately and with an obvious enjoyment, before moving quickly on to the second which was sipped even slower and more deliberately. She was savouring every drop, and the extrovert character lurking beneath the prim exterior was encouraging her to become progressively more giggly and less and less coherent as the alcohol found its way into her blood stream. It was not the behaviour of someone who drank similar amounts as a matter of routine, and her carer wisely suggested that she should perhaps prepare to retire for the night.

The following morning she was full of life and enthusiasm, proclaiming loudly that she hadn't slept so well for many years and singing the praises of the clean Welsh air which she had convinced herself must be responsible.

The same routine was followed on the second night, only this time the carer caved in under enormous pressure from her charge to sanction the purchase of a third brandy. All was well until she got to her feet to totter giggling towards the bar in the vain hope of persuading Irene to let her have another one. She was thoroughly enjoying herself, and alternated between laughing raucously as some dirty joke that wasn't shared with the rest of us, and cackling like an egg bound chicken at anything that took her fancy. Even the most liberal minded of us had to admit that there was overwhelming evidence to suggest that she could with some justification be accused of being a little tipsy.

Every so often she waved her arms at an imagined friend who had just popped into her mind, and they shared the same smutty joke that still retained sufficient impact to bring on the belly laughs again. The arm waving continued as other invisible friends presumably joined in the revelry, and the lady's immaculate appearance took on the dishevelled look of a homeless waif. The beautiful golden brown hair was being tossed around like a cork on a rough sea and made a dramatic bid for freedom by jettisoning its securing pins and bouncing up and down until it tilted precariously over her left ear. Revealed in all its splendid glory, and to the slightly embarrassed amusement of the other guests, was a dignity bashing birds nest of mousy grey coarseness that hadn't seen a brush for months. Its owner was much less restrained than the onlookers and shrieked with laughter as she repeatedly pointed at the absconding wig, just in case there was anyone present who had missed the point.

The lady enjoyed a second night of blissfully sound sleep and emerged in the morning with no apparent after effects or shameful recollections of the previous evenings events. She was once more regaled in her carefully selected finery, immaculately crowned by a head of beautiful golden brown hair.

Before the third evening arrived to cast yet another indelible mark on the new venture, Irene had belatedly learned an important lesson. She had discovered that not all optic measures are the same, and that in future she should without fail make sure that the selected spirits optics, especially for brandy, are for singles and not doubles. A more experienced and wiser barmaid thus not only avoided a substantial deficit in the bar figures but also ensured the future sobriety of our guests. At least we thought she had.

Alcohol was at the root of another lady's problems and the repercussions felt by us and other guests at the time led to a subsequent refusal of accommodation for any unaccompanied people directed to us by the Social Services, unless a written assessment of individual tendencies and potential risks was received prior to the scheduled visit.

The guest in question was a lady of generous proportions who allegedly suffered from every conceivable ailment known to man, and who was prone to taking refuge from her hypochondria in a bottle or two of the hard stuff. Since we weren't inclined to include the restraint of obstreperous drunks in our services, we were only too pleased to agree to the request of her social worker and place the bar out of bounds.

She had evidently been sent down the prohibition road before since this time she had come prepared with her own considerable stock, a fact that was brought sharply to our attention only when the last of it had been consumed in a single binge on the third night. She emerged from her room very much the worse for wear to demand replenishment from our bar and, having her request refused as diplomatically as we were able, disappeared back into her room again muttering something about getting it one way or another. Her choice of words and the manner in which they were delivered suggested that she meant it.

There was little doubt that if she decided to raid the bar by force there wasn't much any of us could do on our own to stop her without jeopardising the safety and best interests of everyone present. We'd all seen her discard the decorative crutches and rush down the passage at ninety miles an hour when dinner was ready, so there were no misconceptions about the fate of anyone standing in her way that night.

Some of our other guests had thought about the problem on our behalf and put forward their own diverse recommendations for consideration. Harry was a true cockney who saw the funny side of everything and suggested that whatever we chose to do he was going to hide behind his wife Nell whose width was exaggerated by her confinement to a wheelchair and provided an excellent shield. Brenda was four feet eleven inches tall in her high heeled shoes and suggested that she would be happy to bring down the transgressor with a rugby tackle as she rushed past, an optimistic proposal that ignored a weight difference of about twelve stones and prompted barely controlled giggles from Graham who assured us that the rapid acceleration of his wheelchair would fill the Centre with the smell of burning rubber. He made no mention of his wife who was also in a wheelchair and would presumably have to fend for herself. Harry picked up on Brenda's idea but insisted that if her strategy was going to be successful we would all have to join hands in order to reach all the way round an ample girth. Graham's imagination went into overdrive and his giggles into hysteria, which only came to an abrupt and messy end when he threw up on our new fitted carpet.

I found myself backed into a rather uncomfortable corner with no choice but to put on a show of calm courage, and confidence in an ability to deal with the matter on my own. Everyone else gratefully retired to the safety of their respective rooms for the duration of the night, and there passed a long lonely eight hours, hopeful that the lady would remain in her drunken stupor but never being quite sure that she would.

They came to collect her the next morning, and disappeared forever in the direction of the gate with my words of admonition ringing loudly in their ears. Harry and Nell are still good friends and regular visitors, and our conversations never fail to somehow include the tragic central figure of that eventful night. Harry had seen it all before, and I strongly suspect that it's the memory of our innocent naivety more than the plight of the unfortunate lady or the hilarious reactions of the other guests that still makes him roar with laughter as soon as the subject is raised.

And so our initiation period was not quite as we had anticipated, nor was it as smooth and tranquil as it could have been, but a few valuable lessons had been learned and we were much wiser and better prepared hosts at the end of it. At least we were now as ready as we could possibly be to welcome the first of the children who had unknowingly spawned the whole idea some five years earlier.

As we waited for their arrival we were feeling like seasoned veterans, confident that, compared to the initiatory traumas, catering for the needs of thirty five children and carers would be a piece of cake. And as it turned out, it was.

Chapter Eighteen

The First Children

The dogs knew they were coming, barking and rushing down the passage to be first on the scene. Then the loud hiss of air brakes signalled their arrival and hours of frantic preparation came to a hurried end. Scurrying bodies appeared from all directions, carrying the tools of the day's labours and hoping that everything was ready.

The air was sweet with the fragrant smells of bathroom sprays and carpet cleaners, the beds were fresh with newly laundered coverings, and polished surfaces in dining room and bar sparkled in anticipation. We were as ready as we were ever going to be.

The coach engine was silenced and an excited face appeared at the doors. It was a young girl's face, about nine or ten years old in boyish jeans and shirt, and with long fair hair hanging untidily around her shoulders. She jumped from the coach, threw us a broad beaming smile, then scrambled back inside again to busy herself with the needs of her friends, sorting out their hand luggage and other bits and pieces that would be urgently needed in the race to organise themselves in the privacy of their rooms.

Jenny was one of the lucky ones, as fit as a flea and twice as lively, with a bubbly personality and motherly instincts, a natural leader respected by her peers. She was just the first of the children we came to know so well over the next few days, children with a variety of different abilities and health problems, but with a common determination to live life to the full and to take advantage of every available opportunity.

The dogs were the first attraction and they made the most of it, wallowing in all the fuss and attention and welcoming the chance to play with children again. Except Candy of course, and she still had more important things to do. She saw the new building as an ideal location to practice her spider hunting skills, and spent a large part of that first day crouching in obscure corners waiting for any unwelcome intruders to show themselves.

The horses helped to set the rural scene outside by congregating around the Centre to find out what was going on and, just in case there was a chance of it being their lucky day, to wait for any tasty treats that may be on offer. And there were, probably more than was good for them.

The three Shetlands had the advantage of being at the same height as the youngsters in wheelchairs, so were a big hit. Lucky and Snuggles made plenty of new friends, but it was Sam who stole the limelight by trying to shake hands with everyone in sight and refusing to take no for an answer until someone got the message and produced the mandatory polo.

Dusky, true to form, used the occasion to quietly sneak in through the fire doors when he thought no one was watching.

The other horses stood quietly to one side, watching the antics of their extrovert cousins and trying to make some sense of all the goings on. Their part in the procedure was yet to come, spending a few hours nearly every day taking youngsters for rides around the mountain land or down towards the stream on the southern boundary.

Even in the first few hours, the children showed an obvious rapport with the animals, with the scenic beauty and with the open spaces. The combination was already working well enough to prove a long established point, and there was little doubt in anyone's mind that the effects would be long lasting. This was what it was all about. These were the people the Centre was built for.

Our feelings were reinforced on the day we took them on the first ever hay ride. It was a simple way of taking them deep into the farm and into a new and natural world of fields and livestock, trees and streams, a world they'd only before seen in pictures. They rode on a long, low trailer towed behind a tractor, with hay bales as seats and more bales as padding for those restricted to wheelchairs. Guard rails that we'd fitted in front and along both sides made sure that they could come to no harm.

The nosy Welsh Black cattle came to say hello and there were squeals of delight and apprehension as they tugged greedily at the hay. Even the stock bull came over to play his part, allowing the brave ones to stroke his massive head. They were introduced to Boofy the Jersey house cow, she was well used to being handled and petted and quickly became a firm favourite. I'd bought her as a silver wedding anniversary present for Irene, so she was more a family pet than anything else.

Yes, I know that a cow is a strange present, especially for such a milestone in our lives. But at least it was different, and ever since Dozy, our first Jersey, had shed her mortal coil a year or so earlier, Irene had repeatedly said she'd like another one. Well now she had one, but I didn't dream at the time that Boofy would prove to be such an asset in such a completely different way.

After a picnic in a field above the stream, we stopped at the sheep shed, encouraging the children to hold the lambs and to take it in turns to feed pets from a baby's bottle. It was a moment we'd looked forward to, and we stood back to watch a natural and rewarding interaction and to listen to the excited chatter.

It was then that we noticed Jenny again. She'd left the others and was sitting alone in the corner of a pen of three weak lambs that had been rejected by their mothers. In her arms was the sickest of the three, a pathetic little creature that was unlikely to survive. She was holding it tightly to her chest, talking quietly to it and rocking gently to and fro. A tiny woolly head was snuggled closely and gratefully under her chin.

That was where she stayed, quietly absorbed in her self appointed role until it was time to go. Then, very reluctantly, a sleepy lamb was gently laid on a bed of soft straw and, staring intently at the ground to hide the tears welling up in her young eyes, Jenny walked silently from the building. Words would have been pointless and we could only guess at her very private thoughts.

The week produced other young stars, children with engaging personalities and unique abilities who left their mark on everyone and were the cause of many a damp eye and lumpy throat. People like John, who loved music and played brilliantly on a little electronic organ using only his knuckles on fingerless hands. And Katie, a little girl with a permanent twinkle in her eyes, whose spontaneous rendition of an extremely long and off key 'Little Donkey' brought loud cheers from her friends - which she took to be a request for an encore so sang the whole thing over again. And then there were the quiet ones, busily making sure that their friends could manage whatever it was they were doing, and helping where they could.

The last day was hectic, with a frantic determination to do all the things there hadn't been time to do during the week. In the afternoon there was a swimming gala, noisy races with staggered starts to compensate for those with more serious physical problems, and then a rush to shower before dinner and to get ready for the planned evening's disco.

There was medication to organise, and some of the children needed gentle massage to boost poor circulation and encourage idle muscles into a degree of activity. The rooms were transformed into hubs of feverish activity, and the shrill noise of excited youngsters found its way into every corner of the building.

From behind one of the closed bedroom doors came a different sound, the pummelling of expert hands on a soft young body, involuntary grunts, and coughing to clear blocked airways as manipulative pressures did their work.

We hadn't thought much about individual health problems, our guests were after all just children, and that's the only way we wanted to think of them. But this was physiotherapy with a difference, harsher and almost brutal, and I remembered hearing that one of the children was waiting for a life saving heart and lung transplant. The stark reality of their lives was forcibly brought home to me, and my own, comparatively minor, problems evaporated into irrelevance.

The coughing stopped and there was a moments silence, broken at last by a low murmur of encouraging words. The door opened and a pale young face prepared to face the world again. It had the drawn look of utter physical exhaustion, but there was also resolve and determination, and not a trace of self pity.

Then Jenny threw me one of her broad beaming smiles and, with a weary toss of her hair, disappeared in the direction of her friends and the ear shattering music of the age. The disco had started, and children were already dancing with all the verve and gusto they could muster, some in wheelchairs pushed and twisted and turned by willing hands, anxious that no one should be left out.

We learnt a lot about courage during that first week, and there was sadness all round when it was time for them to leave. They'd been happy and adventurous, and their stay with us was packed with fun and new experiences. They were a joy to be with and, however hard I try, I can't think of one single thing that went wrong or one single moment that I regret.

The natural beauty of the farm had weaved its wondrous magic, just as it had for us seven years earlier. Blaen Wern itself had spawned the Centre. It was both the instrument of conception and the foundation for its creation. It was both its home and its host. And it was because of children like these that it had all come together. Perhaps that's one of the reasons why we enjoyed every moment of their stay with us. It was a visual confirmation of the fruition of our hopes and plans.

Their departure was an appropriate time to draw a deep breath before the next group arrived, to enjoy a couple of days respite and to look back over the past eleven and a half years since that day in 1977 when we first arrived in Wales with our tiny caravan. There had been very little opportunity in the intervening years for the luxury of rest. I'd led my family along a hard and precarious path, and they had followed without question and with complete trust. If it hadn't worked, the repercussions would have been numerous and severe, and the responsibility would have been mine. Surely now that all the decisions had been vindicated I had every excuse to relax a little, and possibly to even feel a bit pleased with myself. But it didn't happen like that. There was no sense of achievement, only a desperate awareness that it wasn't enough.

Everyone in my family had, in one form or another, made the Centre possible, Irene, Mark, Paul, Nicola, my sister Sybil and her late husband Ray, my Mother, and even in a way my late Father who had died some twenty years before. Together we'd reached the top of the hill and I should have been happy, but all I could see were more, bigger hills. There was so much more that needed to be done that our trifling endeavours seemed insignificant and almost irrelevant. I felt no euphoria, or even satisfaction. If there were feelings at all, they were only of relief that the goal had been reached. It was, of course, the classic anti climax syndrome.

I still see those hills in the distance, and wonder how many more can be climbed before the ravages of time take their toll on available energy, drive and commitment, and before the inevitable need to rely on others becomes imminent.

As I write, the Charitable Trust is beginning to bear fruit, to attract the cash that will enable so many more disabled and sick children to enjoy the freedom and natural therapy of the countryside. The Trust has that responsibility, to look forward, to be wherever sick children's needs may be. That's where its future lies. The Centre has done its work well but its only a building. Its real worth is in the way it will be used in the years to come.

Whatever the future may hold for Blaen Wern, and however much I am forced to delegate, I shall without doubt find a suitable role within the Trust that will keep me in touch with our creations, and allow me to express my views as forcibly as ever.

And perhaps when that day comes I shall find more time to ride or just walk with the dogs, to enjoy the company of an abundance of wild creatures who share Blaen Wern and to marvel at the delicate balance of Nature's work. Perhaps I can even more delight in the raw and unique beauty of our very special home and be satisfied that so much happiness has been given to so many people, simple pleasures that began with personal ideals and a search for our Mighty Oak.